GAVIN WHITFIELD

I0867086

50 MUST-SEE GEOLOGICAL SITES

in South Africa

Rod MacLeod

GSSA Geological Society of South Africa

E OPPENHEIMER & SON SOUTH AFRICA

Council for Geoscience

Contents

Acknowledgements

It is a reality that a book such as this needs sponsorship to get beyond the conceptual stage, so my appreciation must go first to several major benefactors. My thanks go to Ernest Oppenheimer & Son who at the start believed in the concept and who provided a generous donation to get the book going; and a special thank you goes to Duncan MacFadyen, their Manager of Research and Conservation.

No book on a geological topic can be successful without good quality maps, as were supplied by the Council for Geoscience (CGS), at no cost to the author or publisher. My considerable gratitude goes to this organisation and, in particular, to the Spatial Data Management Unit, headed by Ken Wilkinson and Magda Roos, and assisted by Belinda Fourie: they provided excellent geological and digital elevation model maps.

The Geological Society of South Africa also contributed financially, through substantial donations from their Research, Education and Investment Fund, to assist with the field work that was required across the country. Thank you for your support and assistance, which is now bearing fruit, and for helping with the cost of illustrations.

The team at Struik Nature, led by Publishing Manager Pippa Parker, must take full credit for yet another very professional popular science publication. I owe a great deal to the editing skill and patience of Helen de Villiers, and the vision and design ability of Janice Evans. Thank you as well to others in the team, including Colette Stott, who assisted with many photographs, Emsie du Plessis, who undertook the proofreading and indexing, Craig Smith who checked the book for technical accuracy, and others who contributed to the success of the book.

Photographs and illustrations form an essential part of this book; a number of people provided images of geological sites, and are gratefully acknowledged in photo credits alongside their photographs. Most of the illustrations were done independently by James Berrangé who, with considerable enthusiasm, produced attractive pictures with a minimum of supervision. The illustration of palaeo-Karoo life came from the Albany Museum. Illustrations of the Premier kimberlite pipe at Cullinan came from renowned geological artist Maggie Newman. Robert Reisz from the University of Toronto generously made available an exceptional photograph and illustrated reconstruction of an unhatched dinosaur embryo from Golden Gate.

Most of the information provided in this book is available from a wide variety of published literature, much of which is listed in the section marked 'Further reading', and wherein many of the authors are indicated. Nevertheless, special thanks are due to a number of persons who, during the course of the field work, provided either additional information, usually informal, or access to geological sites. These include Pippa Haarhoff, Manager of the West Coast Fossil Park; Doug Cole, Paul Macey, Coenie de Beer and Dave Roberts from the Bellville

office of the CGS; and Greg Botha and Nigel Hicks from the Pietermaritzburg office of the CGS. Wild Coast authority John Costello kindly reviewed the Hole in the Wall geosite. Roger Gibson provided insight into the status of the Vredefort Dome. Assistance on the Barberton Makhonjwa Mountains was given by local scientists Tony Ferrar and Chris Rippon. Stefa Estebeth, the Manager at Sudwala Caves, supplied an unpublished report by speleologist Pedro Boshoff. Greater geological insight into the distribution of the Black Reef came from George Henry. Sian Tiley-Nel, Curator of the Mapungubwe Collection, made available numerous images and Emese Bordy and Adam Bumby supplied geological information on the Mapungubwe area. Richard Wadley and Pat Ericson provided background to the geology of the Waterberg. Access to Makapans Valley came from Mark Howitt at the National Game Breeding Centre in Mokopane.
On the Pilanesberg, Jos Lurie provided a geological overview, and the management of the Hartbeespoort Aerial Cableway provided access to the top of the Magaliesberg. Dominique Stratford kindly reviewed the draft Sterkfontein Caves manuscript. In the Northern Cape, Jock Robey had input on the old Kimberley Mine; and geohydrologist Reinie Meyer produced information on water flow from the Eye of Kuruman. Coenie de Beer from the CGS (Bellville), together with Roger Smith from the Iziko South African Museum, supplied valuable information on the unique Fraserburg Fossil Surface.

DEDICATION

This book is a tribute to all of South Africa's superb geologists, both past and present, who made the fascinating geology of the country understandable; without this invaluable input an explanation of the geological sites would have been impossible. It is also dedicated to my wife Ann, my partner for over 45 years, who supported and encouraged me throughout the process.

Published by Struik Nature (an imprint of Penguin Random House South Africa (Pty) Ltd)
Reg. No. 1953/000441/07
The Estuaries No. 4, Oxbow Crescent (off Century Avenue), Century City, 7441 South Africa
PO Box 1144, Cape Town, 8000 South Africa

Visit **www.penguinrandomhouse.co.za** and join the Struik Nature Club for updates, news, events and special offers.

First published in 2015
10 9 8 7 6 5 4

Publisher: Pippa Parker
Editor: Helen de Villiers
Designer: Janice Evans
Illustrator: James Berrangé
Cartographers: Magda Roos & Belinda Fourie
Consultant: Dr Craig Smith
Proofreader & indexer: Emsie du Plessis

Reproduction by Hirt & Carter Cape (Pty) Ltd
Printed and bound in China by C&C Offset Printing Co., Ltd.

MIX
Paper from responsible sources
FSC
www.fsc.org
FSC® C018179

Penguin Random House is committed to a sustainable future for our business, our readers and our planet. This book is made from Forest Stewardship Council ® certified paper.

Print ISBN 978 1 92057 250 1
ePub ISBN 978 1 92057 251 8
ePDF ISBN 978 1 92057 252 5

All cover images credited alongside their appearance inside the book.

Foreword

South Africa is world renowned for its geology and mineral wealth, with mining playing a huge role in the economy of the country. There are many geological wonders such as numerous diamond-bearing kimberlite pipes; the extraordinary Witwatersrand goldfield, the world's greatest producer of gold; and the platinum-rich Bushveld Complex – all are geological superlatives that have attracted the attention of generations of geologists and world-class mining companies formed to exploit their contained mineral wealth. The importance of South Africa's lesser known geological and landscape treasures, however, which are generally not associated with minerals of economic value, but which are often of outstanding scenic, cultural and scientific value, is frequently overlooked.

In this beautifully illustrated and meticulously researched publication, Gavin Whitfield, a real champion of the geological heritage of South Africa, has highlighted 50 of these outstanding sites throughout the country, and in so doing has made them accessible to the broader public, tourists and geoscientists alike. The geology that has been highlighted covers 3.5 billion years of Earth history and provides an almost continuous record of the development of Earth and of the life it supported. Each site within each of the seven regions of the country has been chosen with care, and photographs of panoramic views, details of interesting rock formations, together with simplified geological maps, geological cross-sections and descriptions make for easy reading and will assist in the understanding of many of our landscapes and geology. A useful adjunct to the geological descriptions of individual sites are lists of things to see and do, including hiking trails, and associated historical and cultural attractions.

The appearance of *50 Must-see Geological Sites* is timeous as it will be of great value for delegates to the 35th International Geological Conference (to be held in Cape Town in 2016) wishing to see some of our outstanding geological attractions before or after the conference. The Board of the 35th IGC Foundation and the Local Organizing Committee for the conference are proud to be associated with this publication, a fitting celebration of South Africa's rich geological heritage.

PROF. RICHARD VILJOEN

Co-President of the 35th International Geological Conference Local Organizing Committee

Preface

This book was conceived several years ago during a 'brain-storming' session between Pippa Parker of Struik Nature and Craig Smith of the Geological Society of South Africa on future geoheritage publications and the idea was passed on to me. The concept of documenting a large number of important geological sites in South Africa was appealing and I agreed to author this guidebook.

But this is part of a trend to bring geology to the greater public, and since the late 1990s a number of earth science publications have emerged in South Africa, and have done well in the bookshops. Pointing the way was *An Introduction to South Africa's Geological and Mining Heritage* (1999) by Morris Viljoen and Uwe Reimold; followed by the highly commended *The Story of Earth and Life* (2005) by Terence McCarthy and Bruce Rubidge, which tells a complex yet fascinating tale of southern Africa's development; *How on Earth?* authored by Terence McCarthy in 2009; and the popular travel guide *Geological Journeys* (2006) authored by Nick Norman and me, which has sold remarkably well, far exceeding the publisher's original expectation. Such successes show that at the popular level there is considerable and growing interest about the nature of the Earth beneath our feet. This is not surprising, bearing in mind numerous recent rumblings of the Earth such as earthquakes, volcanoes and landslides, our increasing environmental awareness and the threat of climate change. Geoscientists are increasingly being asked to present their knowledge to the public and this involves the challenging task of communicating to people with no or little prior knowledge of how the Earth works. One of the main outcomes of this book is that it provides a guide to geological processes and South Africa's geological past.

My travels around South Africa during the course of preparing this book made me very aware that an explanation of the geology is generally missing at almost all sites of natural interest: it is rare to find a magnificent landscape, a spectacular landform or a fold mountain that has an adequate description, be it on-site or in the tourist literature. How many times have you looked at a particular geological feature, such as a mountain range, or a canyon, or a major escarpment, or even a cave, and wondered how it was formed or what it means? This was another spur to write this book.

I have tried to provide information on a large selection of South Africa's best-known geological sites and landscapes, placed in their regional geological context, and to give the observer a good idea of what they are seeing. The book takes in a wide variety of geological and land-forming processes, both large and small. It has been compiled with the interested lay person and visitor in mind and, as far as possible, geological jargon has been kept to a minimum, or at least explained. The book provides illustrated descriptions of geological sites (geosites) that have been selected from a much longer list, based on their geo-educational credentials and accessibility to the public. No-one who has used this book should be able to say that they don't understand how specific geological features were formed.

While effort has also been made to cover as much of South Africa as possible, please remember that there is still a lot more to see and experience!

GAVIN WHITFIELD
Johannesburg, South Africa

Introduction

South Africa's geological heritage

The geological sites (or geosites) described in this book are part of South Africa's incomparable geological heritage. The definition of 'geological sites' has been taken as broadly as possible in order to illustrate specific rock types, geological or structural processes, palaeo-environments, spectacular landscapes or specific landforms; and certain mines are included. Many of the sites are located within protected areas, including World Heritage Sites. Importantly, they are accessible to the public, are relatively easy to visit and can be appreciated by all without undue effort.

The science of geology is commonly defined as 'the study of the origin and nature of the Earth'. Geology includes a wide variety of separate disciplines, including broad-scale mapping of geological strata to create geological maps, computer-assisted interpretation of aerial photos and satellite images, prospecting and exploration for valuable resources, environmental and mining studies, the study of fossils, and detailed microscopic and chemical investigations of rocks and minerals. Geomorphology is the study of the surface forms of the Earth or of landscapes, in particular the processes that relate to the origin and evolution of landforms. There is a fundamental relationship between the underlying geology and the landscape, and the landforms that have been created at all levels.

Southern Africa's vast span of geological time provides an opportunity to see the country's unique geological record, from nearly 3,600 million years ago to 'Recent' times, and to get insight into how the Earth and its infinite variety of life forms evolved over the immensity of geological time. Some of southern Africa's most impressive large-scale geoheritage attractions, from oldest to youngest, include:

- **Barberton Mountain Land:** One of the best preserved volcanic, sedimentary and granitic micro-continents that grew and eventually united to form the Kaapvaal Craton. Some of the Barberton rocks provide evidence of very primitive, single-cell microbes that lived around 3,400 million years ago in a primordial sea.

Site BM7 on the new Barberton Makhonjwa Geotrail provides representative mounted specimens of all rock formations seen on the 37km-long road trail.

- **Witwatersrand Basin:** A little less than 3,000 million years ago, after the Kaapvaal Craton had enlarged and stabilised, down-warping of the craton took place and a large inland sea formed. The resulting basin began to fill with sedimentary material, including the gold-bearing quartz pebble conglomerate reefs of the Witwatersrand goldfield.
- **Transvaal Basin:** Around 2,650 million years ago this large sedimentary depository started forming on top of and alongside the earlier cratonic basins, which by now had become hard rock formations; and vast marine beds of photosynthesising cyanobacteria developed, which produced

copious algal-generated oxygen that was released into the sea water, resulting in the deposition of South Africa's immense sedimentary iron and, later, manganese deposits. Life on Earth, albeit marine, had never before existed on such a grand scale. These fossil algae are now preserved as thick layers of dolostone, also called 'dolomite'.

- **Bushveld Complex:** Over a relatively short period of geological time, around 2,060 million years ago, the Kaapvaal Craton was massively intruded by magmas of deep-seated origin to create what is by far the world's largest layered igneous intrusion. South Africa thus became host to the world's biggest resources of platinum, chromium and vanadium, and the pre-eminent producer of these commodities for the foreseeable future.
- **Vredefort meteorite impact:** Not long after this, Earth was struck by an enormous meteorite, now leaving only its roots as the central uplift – caused by the rebound of the underlying rock – of the once-enigmatic Vredefort Dome. This has now been identified as the world's largest and oldest known meteorite impact structure, dating from 2,020 million years ago, and the impact is considered to have resulted in the Earth's largest single release of energy. This catastrophic event significantly influenced the distribution of the rich Witwatersrand gold reefs that became, for a time, the backbone of South Africa's economy. These conglomerates were to yield the largest gold haul of all time: more than 50,000 tons of gold has already been mined, and a lot is still unmined.

And, at the opposite end of the time-scale, only around 200,000 years ago, another much smaller meteorite struck, creating the Tswaing Crater – one of the youngest, best preserved and most accessible bowl-shaped meteorite impact craters on Earth.

- **Waterberg and Soutpansberg Basins:** In these basins, the world's earliest red beds formed. Rich in iron oxide, they reveal that, by at least around 1,900 million years ago, oxygen was freely present in the atmosphere, a situation that eventually enabled life to exist on land and changed the Earth's future forever.
- **Cape Fold Belt:** The magnificent mountain ranges of the southern and Western Cape owe their existence to the movement of crustal plates far south of the Gondwana supercontinent around 300 million years ago. A subduction zone developed along the southern margin of Gondwana as oceanic crust moved under the continent. Sedimentary rocks previously deposited in the Cape Basin became compressed and formed the elongated fold belt. The result was a mountain range of Himalayan proportions and, although it's now considerably worn down, the passes and poorts provide a marvellous geological spectacle.

More than 500 million years ago, and prior to deposition of the Cape sedimentary rocks, an earlier mountain-building event caused large-scale crustal deformation and resulted in the intrusion of numerous batholiths and plutons of Cape granite.

Folded Table Mountain quartzitic sandstone in the little-known Seweweekspoort, which cuts through the Great Swartberg near Ladismith. In the distance is the 2,325m-high Seweweeksberg, the highest peak in the range.

- **Karoo Basin:** The rising mountain ranges along the southern Cape coast resulted in a vast sedimentary basin developing further to the north. Some primitive life

Coloured Beaufort shales of the Karoo Basin in Teekloof Pass, an area well known for its reptile fossils and intruded by many Karoo dolerite sills, one seen capping the sedimentary strata at the top left.

that had evolved under water gradually moved onto the land, plants flourished and around 250 million years ago the great alluvial flood plains saw the progressive development of a variety of prehistoric reptiles and, eventually, the ancestors of early mammals. But by about 190 million years ago the basin was reduced to a sandy desert, seen today as the Clarens sandstone.

- **Karoo Igneous Province:** As a forerunner to the break-up of Gondwana, the Earth's crust started to crack and, around 180 million years ago, basalt magma rose up and flooded across most of southern Africa, making one of the world's great continental basalt provinces. At the same time, basalt magma intruded existing Karoo sedimentary formations to form dolerite sills and dykes, ultimately giving rise to the region's iconic Karoo scenery.
- **Kimberlites:** At various times during the geological past there were explosive eruptions from kimberlite volcanoes, the primary source of all natural diamonds. Kimberlite was first discovered in South Africa in 1871 and the country was for many years the world's largest producer of diamonds. In 1905 the Premier kimberlite pipe produced the famous Cullinan Diamond, the largest gem diamond ever found, and the same pipe is still producing world-class gems.
- **Great Escarpment:** South Africa is renowned for its magnificent landscapes, and much of the spectacular scenery can be traced back to the break-up of the Gondwana supercontinent, which started more than 140 million years ago. This resulted in large-scale rifting of the continental margins, the creation of new oceans, uplift of southern Africa, down-cutting by coastal rivers and massive denudation.

Part of the Blyde River Canyon, viewed looking west, showing the Great Eastern Escarpment being cut back by erosion. Capping the high flat ground in the distance on the right is the thin blanket of Black Reef quartzite.

- **Human evolution:** The widespread Transvaal dolostone beds mentioned above are riddled with cave systems that developed very much later, and it is a strange coincidence that these caves witnessed and protected the evolutionary path of mankind, starting with our distant apeman ancestors around 3 million years ago. Many renowned fossils from this time have been recovered from numerous cave sites associated with the world-famous Cradle of Humankind World Heritage Site.

About this book

- This book showcases 50 of the best geological sites in South Africa: some are diamond mines, others fossil sites and many are landscapes that will be known to you in passing. Their locations are shown on the maps on pp.12 and 13 (and see note alongside).
- Naturally, the selection of geological sites has been somewhat subjective and is by no means inclusive; many other interesting sites, tucked away in remote and rugged places, are omitted.
- While the main aim of this book is to describe and interpret places of geological interest across South Africa, the 'Introduction' offers readers an overview of the geology and its companion science of geomorphology.
- A Glossary (pp.312–314) provides explanations of many technical terms.
- 'Further reading' (p.315) provides a list of recommended publications, giving access to more detailed information.

TIP Most geological and landscape information can best be seen from above. Google Earth allows you to 'fly' anywhere on the planet to view combined satellite imagery and aerial photography in amazing topographic detail and obtain GPS co-ordinates. All geosites presented here can be viewed on-line at various scales – a highly recommended tool for the 'arm-chair' geoscientist.

A note about maps

Country-wide landscape and geological maps (pp.12–13), compiled using digital terrain technology and simplified geology, show the entire region and indicate the distribution of all featured geosites (numbered **1–50**).

Six of the regions (all except Northern Cape, where the geosites are too widely distributed) have larger-scale, more detailed landscape and geological maps that also locate more precisely the featured sites. In each instance, the twinned maps emphasise the strong connection between the underlying geology and the formation landscape.

A location map shows key routes and towns in the vicinity of each site: GPS co-ordinates pin-point the feature, or closest vehicular approach.

The Council for Geoscience (CGS) is able to supply detailed information and more comprehensive regional maps.

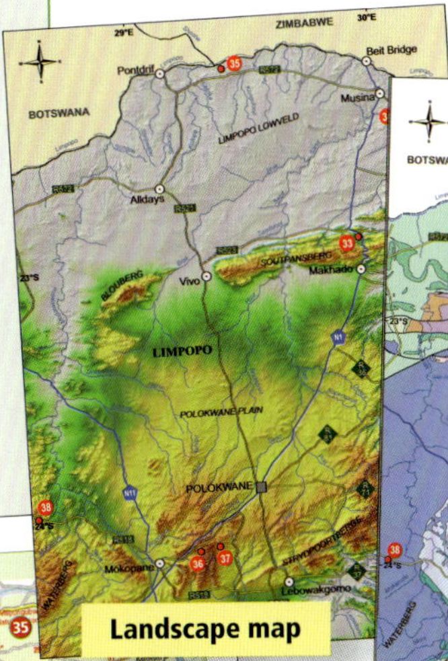
Landscape map

A key to the simplified regional geological maps appears on the inside covers, front and back.

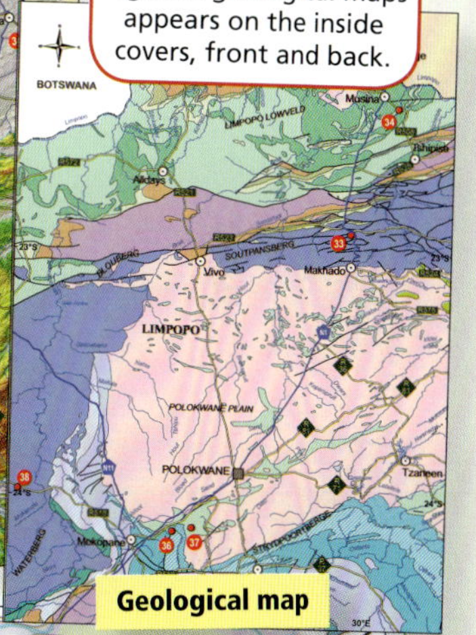
Geological map

Site location map

Figure 1: Landscape map of South Africa and Lesotho, derived using digital elevation technology and showing the locations of all featured geological sites.

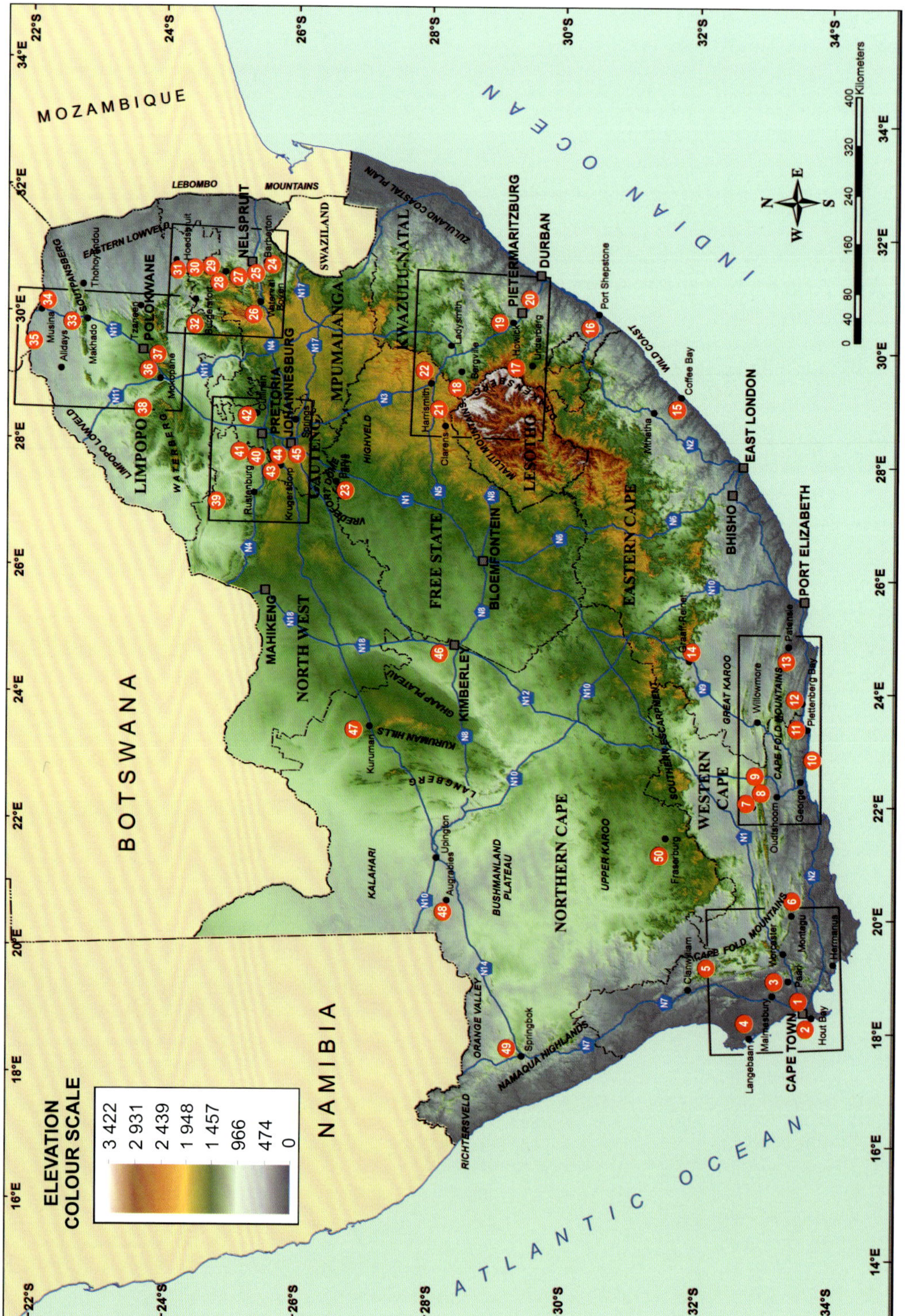

Figure 2: Simplified geological map of South Africa, Lesotho and Swaziland. Boxes delineate six of the seven regions discussed, although some geosites fall outside of the more detailed regional maps at the start of sections, and Northern Cape is without a regional map as the sites are too widely dispersed.

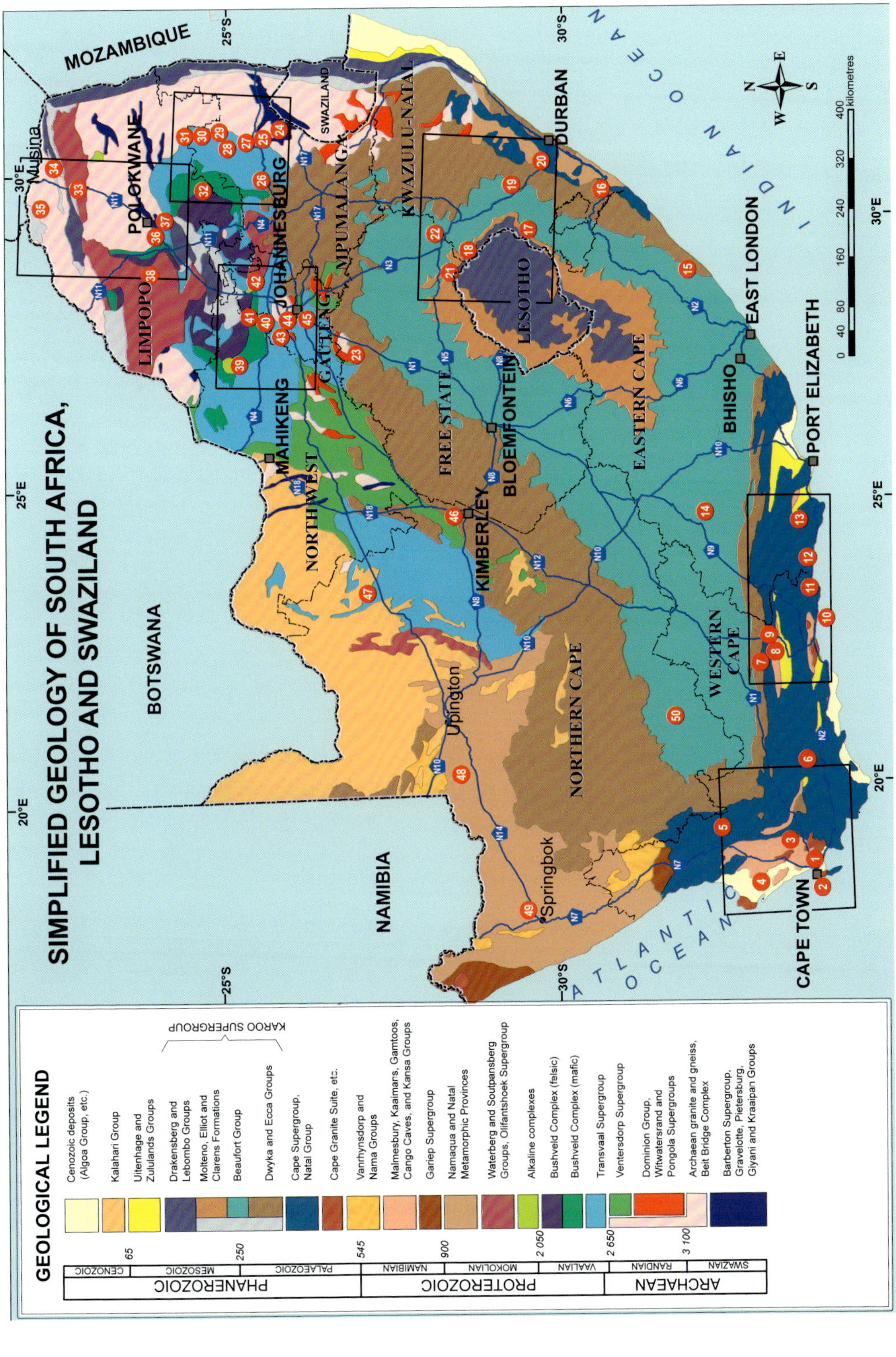

What rocks are made of

Rocks are aggregates of one or several common minerals, and are defined by these varying mineral proportions and by grain size. Minerals are naturally occurring inorganic compounds and often have complex chemical compositions. They have characteristic physical properties and crystal structures, and all minerals have approved names. There are about 4,000 of them, but the majority are uncommon to rare. The most common rock-forming and accessory minerals are listed in **Table 1** below.

Table 1

ROCK-FORMING MINERALS			
Name and varieties	**Occurrence**	**Essential chemistry***	**Colour**
Quartz and other silica varieties Also found as **rock crystal, agate, chalcedony, chert** and **jasper**	Found in igneous and metamorphic rocks of granitic composition; in quartz veins; and in most sedimentary rocks; the main component of sandstone	SiO_2 (also called 'silica') and the building block of all silicate minerals	Colourless to white, many coloured varieties
Alkali Feldspar Group Ranges from **albite** (sodium-rich) to **orthoclase** or **microcline** (potassium-rich)	Alkali feldspars are among the most common minerals; a major constituent of igneous and metamorphic rocks	SiO_2 (silica) and Al_2O_3 (alumina) with variable Na and K	White to pinkish, even reddish
Plagioclase Feldspar Group From **albite** (sodium-rich) to **anorthite** (calcium-rich)	Plagioclase feldspars are common in intermediate and mafic igneous rocks and in many metamorphic rocks	SiO_2 (silica) and Al_2O_3 (alumina) with variable Ca and Na	Whitish to pinkish
Mica Group Ranges from **muscovite** (potassium-rich) to **biotite** (magnesium- and iron-rich)	Primary minerals in igneous and metamorphic rocks. Muscovite is found as flakes in sedimentary rocks. Common in pegmatites	Series of very platy hydrated K-, Mg- and Fe-bearing alumino-silicate minerals	Almost colourless to golden or dark brown
Pyroxene Group Includes **augite** (common) and **enstatite** (magnesium-rich) and **diopside** (calcium-rich)	Common in high-temperature gabbros and dolerite, and ultramafic rocks of mantle origin	Complex alumino-silicate series with variable Ca, Mg, Fe and Na	Dark brown, green or yellowish crystals
Amphibole Group Includes **hornblende** (common), **tremolite** and fibrous **riebeckite** (blue asbestos)	Common in intermediate and mafic igneous rocks and alkaline rocks. Very common in metamorphic amphibolite	Complex hydrated alumino-silicate series with variable Ca, Mg, Fe and Na	Dark brown or green; often altered to chlorite
Olivine Series End-members are **fosterite** and **fayalite**; gem quality is called '**peridot**'	Common in high-temperature ultramafic rocks (peridotite, komatiite) and in some mafic and metamorphic rocks	Series ranging from magnesium-rich Mg_2SiO_4 to iron-rich Fe_2SiO_4	Green to brownish; alters to serpentine.

Name and varieties	Occurrence	Essential chemistry*	Colour
Epidote Group Includes **zoisite**; the rare blue-mauve form is called **'tanzanite'**	Widespread metamorphic and hydrothermal mineral found mostly in altered igneous rocks and in veins	Complex hydrated alumino-silicate series with variable Ca and Fe	Pistachio or dark green, also orange to reddish
Garnet Group Numerous species including **almandine, andradite, grossular** and **pyrope**	Widespread in high-grade metamorphic schist, gneiss and marble; pyrope is found in kimberlite	Complex alumino-silicate series with variable Fe, Ca and Mg	Dark red to brown to blackish, or yellowish to orange
Chlorite Group A general name for several mineral species, the most common being **clinochlore**	Widespread and formed by breakdown of primary alumino-silicates. Also a low-grade metamorphic mineral and found in fine-grained sedimentary rocks	Clay-like, complex hydrated alumino-silicate series with variable Fe and Mg	Usually greenish, fine-grained and earthy
Kaolinite Common in a group of secondary clay minerals, often massive; used to make pottery	Formed from breakdown of feldspar minerals by weathering or alteration, and in fine-grained sedimentary rocks	Complex hydrated K-rich alumino-silicate	Usually whitish or creamish, fine-grained and earthy
Calcite and Dolomite Common carbonate mineral species, found as discrete grains, as mineral cement and as rock-making minerals	Sedimentary limestone and dolostone, found as matrix cement, as secondary calcrete, and as travertine. Also in metamorphic marble	Carbonate minerals, $CaCO_3$ or $CaMg(CO_3)_2$	Usually whitish (calcite) to greyish (dolomite)
Iron ores Includes **magnetite, hematite** and **limonite** (iron-rich); **ilmenite** (titanium-rich) and **chromite** (chromium-rich)	Most within mafic igneous rocks, either as concentrated layers or accessory minerals. Hematite and limonite are more oxidised forms, usually secondary in nature	All Fe-oxide minerals, which may also contain Ti, V and Cr	Black to grey metallic minerals; hematite is reddish; limonite is 'rusty'
Sulphide minerals Carriers of economic metals such as copper, zinc, lead, nickel plus gold and platinum metals	In many rock types as accessory or rare minerals. They form under reducing conditions, and oxidise rapidly near surface	Commonly include **pyrite** and **marcasite** (both FeS_2) and **pyrrhotite** ($Fe_{1-x}S$)	Brassy or bronzy, metallic appearance

***Elements mentioned in Table 1**

Ca – calcium
Cr – chromium
Fe – iron
K – potassium
Mg – magnesium
Na – sodium
O – oxygen
S – sulphur
Al_2O_3 – alumina molecule
CO_2 – carbon dioxide
Si – silicon
SiO_2 – silica molecule
Ti – titanium
V – vanadium
Hydrated – contains water, H_2O

Rock types and their classification

Over the years geologists have delighted in giving names to different types of rock, often naming them after the place where they were first found. The result is that there are hundreds of rock names, but these can be resolved to fewer than 50, and divided into three fundamental rock groups: **igneous, sedimentary** and **metamorphic**.

Igneous rocks

Deep within the Earth, magma (molten rock) is generated and rises through the crust. Before reaching the surface, and sometimes at great depth, it may slowly cool and crystallise into solid rock; rocks formed in this way are known as intrusions, and form plutons (igneous rock bodies), dykes and sills (vertical and horizontal sheets). If magma reaches the Earth's surface, the red-hot molten lava erupts into the atmosphere, or sometimes under water, cools very quickly and forms extrusions. Volcanic eruptions form rocks of this igneous type. Igneous rocks were the first to form on primordial Earth more than 4,000 million years ago, and many different types of such rock have evolved over geological time. As shown in **Table 2**, the classification of igneous rocks depends on both rock chemistry (shown by mineral composition) and grain size, which is linked to the depth at which it solidified.

Grain size: Igneous rocks are described as being coarse-grained, medium-grained or fine-grained to glassy, and the size of mineral crystals that form depends on the time it took the magma to crystallise. The rate of cooling depends on their depth within the crust: deep, slow-cooling magma results in a coarse-grained rock (e.g. granite plutons); at a shallow depth the rocks formed have cooled somewhat faster and are medium-grained (e.g. dolerite sills); and magma that reaches the surface cools quickly and results in fine-grained rocks (e.g. basalt lava).

Mineral composition: Depending on the chemical composition of the magma, a range of different minerals will form, resulting in

Table 2

CLASSIFICATION OF COMMON IGNEOUS ROCKS

Igneous rocks are classified into three categories according to grain size (average size of crystals) and four classes depending on mineral composition, mainly silica content.

	Felsic High silica, light-coloured	**Intermediate** Moderate silica, medium-coloured	**Mafic** Low silica, dark-coloured	**Ultramafic** Very low silica, dark-coloured
Common rock-forming minerals	Quartz, alkali feldspar and micas	Alkali feldspar, amphiboles	Plagioclase feldspar, amphiboles and pyroxenes	Olivine, pyroxenes and serpentine
Volcanic Fine-grained to glassy (<1mm)	Rhyolite, Felsite and Trachyte	Andesite	Basalt	Komatiite (rare)
Hypabyssal Medium-grained (1–5mm)	Microgranite and Granophyre	Microdiorite	Dolerite and Diabase	Kimberlite and Carbonatite – also as volcanic phases
Plutonic Coarse-grained (>5mm)	Granite, Granodiorite and Pegmatite	Diorite, Syenite and Foyaite (rare)	Gabbro, Norite and Anorthosite	Pyroxenite, Peridotite and Dunite

Shown here in the Teekloof Pass is a thick, northward-dipping sill of Karoo dolerite, intruded into horizontal blueish-grey shales of the Beaufort Group. These extensive sills form the most resistant part of the Great Escarpment in the Nuweveld Mountains south of Fraserburg. Notice the columnar jointing on the upper left side.

A much younger mafic (dolerite) dyke several metres thick cuts cleanly across very much older highly deformed gneiss in the valley of the dry Sand River near Musina.

different rock types, and these minerals will also start to crystallise at different temperatures. Rocks with high silica content are called 'felsic' (rich in feldspar and quartz), those with less silica are called 'intermediate', those with lower silica are called 'mafic' (rich in magnesium and iron) and those with very low silica are termed 'ultramafic' rocks.

Sedimentary rocks

The Earth's crust is unstable and dynamic, and erosion of the early igneous rocks, transport of detritus by streams and rivers, and deposition of resulting sediment into the oceans led to the formation of the first sedimentary rocks. Gradually, thick deposits of sedimentary material that accumulated on continental shelves, in ocean basins or troughs became consolidated into hard rock. Over geological time deposited sediment became buried, compacted and cemented, and eventually formed hard rock through a process called 'lithification', which takes millions of years.

Not all sediment is deposited in deep water; for example, much of the sediment deposited in the slowly sinking Karoo Basin is alluvial – deposited by periodic flooding of major rivers that crossed vast flood plains. In the same way that the deposits in ocean basins became lithified, so did the thickly accumulated alluvial sediment.

Classes of sedimentary rock: Three different classes of sedimentary rock are formed: clastic, chemical and organic, and these are classified primarily by grain size and their mineral composition.

Clastic rocks (**Table 3**) are formed from sedimentary detritus deposited as thick or thin beds or strata, under a wide variety of conditions: on the floor of seas and oceans, on growing river deltas entering the sea (deltaic sediments), along river flood plains (fluvial sediments), on offshore bars and beaches (shallow marine sediments) and in lakes and estuaries (lacustrine and estuarine sediments). Volcanic eruptions can also produce sedimentary deposits, called 'agglomerates' (coarse, and found near the volcanic source) or 'tuffs' (fine volcanic ash) resulting from atmospheric fall-out, either into water or on land. Tillite (also called 'diamictite') forms unsorted beds derived from rock debris deposited by glaciers and ice sheets. Strong wind also forms fine-grained aeolian sedimentary deposits that are essentially ancient sand dunes.

Less common are sedimentary rocks of chemical or organic origin, formed by chemical deposition under more restrictive environmental conditions, e.g. limestone, banded iron formation or chert, while organic-rich rocks are represented by oil shale, coal or some types of biogenic limestone (**Table 4**).

Table 3

CLASSIFICATION OF SEDIMENTARY ROCKS OF CLASTIC ORIGIN			
Clastic rocks are categorised according to grain size, a feature of how deposition took place, ranging from extremely coarse-grained to very fine-grained.			
Rock type	**Consists of**	**Grain or particle size**	**General rock description**
Conglomerate	Rock fragments, boulders and pebbles	Very coarse, >2.5mm	Very coarse and heterogeneous forming a concrete-like mixture. The gold-bearing Witwatersrand conglomerates are well known
Sandstone and grit	Sand, usually quartz-rich, also feldspar	Coarse to medium, 2.5–0.05mm	Generally made of quartz sand but may contain other resistant minerals. Usually shows bedding and cross-bedding
Siltstone	Silt, usually quartz-rich, also feldspar	Fine, 0.05–0.005mm	Transitional grain size between sandstone and mudstone, may be quartz-rich and contain less clay
Mudstone and shale	Mud and clay	Very fine, <0.005mm	Deposited in either deep water or as river 'overbank' deposits. Shale is well layered, mudstone is more massive
Agglomerate and tuff	Erupted volcanic debris and ash	Variable, very coarse to very fine	Made from volcanic debris that was explosively ejected. May have been deposited on land or in water
Tillite (diamictite)	Unsorted rock debris moved by ice sheets and glaciers	Very variable	Essentially a type of conglomerate formed by clastic deposition from melting ice

Table 4

CLASSIFICATION OF SEDIMENTARY ROCKS OF CHEMICAL AND ORGANIC ORIGIN			
Chemical sediments are usually very finely layered as a result of slow deposition under very calm conditions; organic rocks, such as coal beds or limestone, are variable.			
Rock type	**Consists of**	**Grain or particle size**	**General rock description**
Limestone and dolostone	Carbonate mud, algal layers or stromatolites; or compacted coral	Fine to medium; some may contain shells	Usually thick deposits of organically derived carbonate. Dolostone is formed by partial replacement by $CaMg(CO_3)_2$. Some limestones are from coral reefs
Chert	Precipitated or replaced crypto-crystalline silica	Very hard, flinty rock, extremely fine-grained	Usually associated with beds of dolostone and banded ironstone. It is dense and resists weathering
Banded ironstone	Precipitated iron oxides, carbonate and silica	Very fine unless later recrystallised	Extremely well banded, usually with alternating layers of iron oxides, chert and jasper
Coal and peat	Carbonised plant material	Fine, semi-crystalline in part	Plant debris, often with minor impurities, accumulated under swampy, oxygen-free conditions
Black shale and oil shale	Accumulated fine organic matter and other fine sediment	Fine-grained, may contain carbon-rich layers	Very dark-coloured, shale-like, consists of fine organic material deposited under oxygen-free conditions
Calcrete and silcrete	Re-deposited carbonate or silica found near surface	Variable, from hard, fine to crumbly	'Younger' deposits of secondary origin capping older land surfaces. Cementing material is derived from underlying rocks

The rock formations at Brenton Beach near Knysna were originally sand dunes deposited by strong coastal wind and later consolidated into hard dune rock. The rugged outcrops are now again being worn down into sand by wind and sea erosion.

Grain size: The grain size of clastic sediment (much more common than rocks of chemical or organic origin) depends on the energy of the water that transported it or reworked it, and ranges from very coarse gravels to fine muds. Coarse gravels that form pebble beds, and ultimately conglomerates, are deposited by fast-flowing rivers and streams, usually when in flood, and often reworked along beaches. Sand is carried further out to sea, coarser sand and grit being deposited first, the finer sand being deposited in deeper water. Strong currents along the coast and advancing or retreating sea levels that result in prograding or retrograding conditions constantly rework the deposits. Silt-size particles are deposited still further away in deeper water, eventually forming siltstone, followed even further away by clay-size particles that ultimately form shale.

Metamorphic rocks

Metamorphic rocks (**Table 5**) are either igneous or sedimentary rocks that have been transformed, physically and/or chemically, under extreme conditions. In a region that is geologically as old and complex as southern Africa it is not surprising that metamorphic rocks make up an important part of the geology. These rocks have formed in places where large-scale crustal processes related to plate tectonics have involved the deep burial and compression of rock formations, or where heat and pressure within the Earth's crust (lithosphere) have radically changed the original rock formations. The metamorphic process takes several different forms: rocks can be transformed by being subjected to intense heat from an igneous intrusion (thermal or contact metamorphism); by being compressed by strong folding on a large scale (dynamic metamorphism); or by being buried deeply (burial or regional metamorphism).

Table 5

CLASSIFICATION OF ROCKS OF METAMORPHIC ORIGIN

Metamorphic rocks originate from earlier rock types that have been substantially altered and changed. Metamorphic intensity results in rocks varying from low-grade to high-grade (not shown here), characterised by new metamorphic minerals and/or new textures.

Metamorphic rock	Original rock	Texture and grain size	Produced by
Amphibolite	Mafic lava and intrusive, or some sedimentary rocks	Medium-grained, usually foliated or gneissic	Alteration of pre-existing mafic rocks or some iron- and lime-rich sediments, under high-grade metamorphism
Gneiss	Granite or granodiorite	Streaky, banded and coarse-grained	Regional metamorphism of granitic rocks, or highly altered sedimentary formations
Hornfels	Mudstone or shale	Hard, fine-grained and locally baked	Contact (heat) metamorphism of fine-grained sedimentary rock, usually of local occurrence
Marble	Limestone	Usually massive to granular	Metamorphic recrystallisation of pre-existing rock, usually made of calcite grains plus impurities
Migmatite	Gneiss and granite	Strongly veined, coarse-grained and streaky	Large-scale partial melting of banded gneiss and invasion by granitic veins, at high temperature
Phyllite and schist	Fine-grained clastic or volcanic rock	Medium- to fine-grained, platy texture	Low- to medium-grade metamorphism under strong folding, typically contains flaky micas or talc
Quartzite	Sandstone	Medium-grained, granular, usually shows bedding	Regional metamorphism of clean sandstone beds; original quartz grains have become recrystallised
Slate	Shale or mudstone	Fine-grained, very good platy cleavage	Strong folding and low-grade metamorphism of shale, to form very well-foliated rock, with platy layers

How sedimentary geological formations are made

'Supergroups' generally cover a large area, while 'groups' may cover a more limited area. Each is formed under unique sedimentary (or sometimes volcanic) depositional conditions and can be dated to a particular geological period. The nature of the geological sequence that forms is a reflection of the type and size of basin or trough into which eroded rock or volcanic material has accumulated over an extended period of geological time.

At some stage crustal conditions around basins change, ending the deposition of sediments; and if sedimentary deposits in a basin are buried deep enough, they will eventually harden to form sedimentary rock by a process called 'lithification' – and so make up new formations that may ultimately become part of a new group or even supergroup.

Later, other forces in the crust may come into action: over time, the now lithified sedimentary basin may be slightly tilted, or involved in a major episode of folding or thrusting, or become faulted or rifted, or be intruded by igneous rocks, and the original formations may become metamorphosed and perhaps extremely deformed. After the elapse of considerable geological time, and possible crustal uplift, erosion and planation, conditions may result in another sedimentary basin being formed, and a new group or supergroup being deposited on top of the older one. It is the task of geologists to recognise these differences and categorise rocks in their correct stratigraphic and structural framework.

Geological formations

The recording and mapping of geological formations follows a standard, internationally accepted system of classification based on their geological characteristics. For all sedimentary and most metamorphic rocks this hierarchical system starts at the level of the supergroup and cascades downwards into groups (and sometimes subgroups), formations, members and finally beds. The system is underpinned by the Geological Timescale (**Table 6**, see p.24), which puts each geological sequence into its correct time context. This branch of geology is known as stratigraphy and in South Africa is managed by the South African Committee for Stratigraphy. All geological formations in South Africa have their place within a country-wide and well-documented scheme of classification, and are relatively fixed in geological time.

It is easiest to understand the naming process by means of a South African example that is familiar to most. The **Karoo Supergroup**, which covers about 65% of the country, is made up of six major 'groups', which are, from the bottom upwards, known as the **Dwyka**, **Ecca**, **Beaufort**, **Stormberg**, **Drakensberg** and **Lebombo Groups**, the latter two being of volcanic origin. These six groups were deposited within the enormous Karoo Basin over a period of about 120 million years, during which time the basin slowly subsided and shrank in area. Each group is made up of several distinct 'formations'; for example, the **Stormberg Group** comprises the **Molteno**, **Elliot** and **Clarens Formations**, which overlie each other and were formed in the same shrinking basin under different depositional conditions, and are locally subdivided even further.

This same principle is followed for all geological sequences in South Africa, from supergroup downwards, and they are usually named according to geographic place names or areas where the geology is representative of a particular group or formation. For example, the picturesque village of Clarens lends its name to the surrounding very characteristic sandstone, the **Clarens Formation**, and the same name is used for the identical formation that may be hundreds of kilometres away.

Intrusive igneous rocks, such as the well-known intrusion of the Bushveld, are collectively referred to as a 'complex', or in part as a 'suite', for example the **Rustenburg Layered Suite**. Similarly, the numerous granitic plutons in the Western Cape are collectively referred to as the **Cape Granite Suite**. A large number of intrusive dykes may be called a 'swarm', usually with a descriptive name such as the **Pilanesberg Dyke Swarm**; and in the case of numerous related intrusives, a 'province' – for example, the **Karoo Igneous Province**. Large areas of strongly metamorphosed and deformed rocks are also referred to as 'provinces', such as the **Namaqualand Metamorphic Province**; or if there are substantial differences in parts, as 'terrains', for example the **Bushmanland Terrain** of Namaqualand.

The simplified geology of South Africa is illustrated in **Figure 2** on p.13, showing the geology down to group level – generally good enough to get an understanding of the regional picture. This is to be read with the Stratigraphic Key on the inside back cover, which lists all the major groups in order of age, ranging from the oldest rocks of the greenstone belts, at about 3,570 million years, to Recent formations that are less than 1 million years old.

The entire east-dipping Bokkeveld Group is displayed in this classic view taken in the eastern Cederberg near Grootrivierhoogte. The group forms a continuous sedimentary depositional sequence of 11 distinct and alternating formations, six of mudstone and siltstone, and five thinner layers of sandstone. These units where deposited within the same ocean basin but under a variety of conditions, and over a long period of time within the Devonian. The base (far left) is the underlying Table Mountain Group and the top-most sandstone ridge is the start of the overlying Witteberg Group.

Geological time

One of the most challenging geological concepts is understanding the immensity of 'geological time', as well as mastering the confusing and unfamiliar terminology that portrays it. We can perhaps appreciate events that took place thousands, or even tens of thousands of years ago, and maybe even a million years does not stretch our imagination too far. But to think in terms of many millions – or even billions – of years is incredibly challenging. Nevertheless 'geological time' is an unalterable fact of science and one we have to accept.

Today, geological time (also called 'deep time') starts with the Hadean Eon, lasting from the creation of the Earth around 4,600 million years ago to the start of the Archaean Eon 4,000 million years ago. Geoscientists continue to push that number back towards a time when the Earth was still a slowly cooling molten mass. At present the oldest dated rock formations come from Canada, either the Acasta gneiss of North West Territory (dated at 4,031 million years) or the Nuvvuagittug greenstone of northern Quebec (4,280 million years), but the latter is still being debated. The oldest measured geological sample (not a rock) is a minute zircon crystal that comes from the Yilgarn Craton in Western Australia; it has been dated at 4,404 million years, and is thought to represent the probable age of the Earth's first crust.

Over the years earth scientists have constructed a workable, albeit extremely detailed Geological Timescale, which is regularly updated and refined by the International Commission on Stratigraphy. This timescale is divided into four Eons, 10 Eras, 12 post-Cambrian Periods and numerous Epochs and Ages, but only those intimately involved are able to comprehend the system fully. In the geological site descriptions that follow, only Periods (for example Cretaceous or Jurassic) will generally be used for Cambrian

Dating and naming of rock formations

In the early 19th century scientists in Europe began systematically to study the nature and order of geological sequences. Age differences began to be recorded of the different layers of fossil-bearing sedimentary rocks, keeping in mind the Law of Superposition – that sedimentary layers are deposited in a time sequence, with the oldest layer at the bottom and the youngest layer at the top. At that stage geological time could only be measured in relative terms, i.e. one rock formation was seen to be older than another. Periodic changes in the type and number of fossils in the various strata resulted in different geological formations being identified and named, the names being derived from local places, and these are generally still in use today. For example, the name of the Cambrian Period comes from 'Cumbria', the ancient name for Wales, where rocks of this age were first studied; similarly, the name for the Jurassic Period comes from the Jura Mountains that lie between France and Switzerland. At that stage fossils were the most important and reliable indicator of geological age and correlation, but in South Africa this does not work for Proterozoic rocks, formed at a time before hard-shelled life existed.

It was only in the early 20th century that the discovery of radioactive dating methods changed the way rocks could be dated. This method uses the principle of naturally occurring radio-isotopes that decay at known, fixed rates over geological time. Provided that the right minerals exist in a rock, and that later geological events have not re-set its internal geological clock, it is possible to measure absolute ages with great precision.

and younger time, while Eons (for example the Proterozoic or Archaean) will be used for the much more distant Proterozoic time. The most up-to-date information on the Geological Timescale, considerably simplified, is shown in **Table 6**. Also shown in the context of geological time are major events that have affected South Africa's geology, including five major extinctions, and a probable sixth now in progress.

Table 6

NB: NOT TO SCALE

GEOLOGICAL TIMESCALE AND MAJOR EARTH EVENTS IN SOUTHERN AFRICA'S PAST

EON	ERA	PERIOD		EPOCH	LOWER AGE (MY)	SOME MAJOR EARTH EVENTS IN SOUTHERN AFRICA – MILLIONS OF YEARS AGO (MY)
PHANEROZOIC	CENOZOIC	QUATERNARY		HOLOCENE	RECENT	6th MASS EXTINCTION NOW IN PROGRESS
				PLEISTOCENE	2.6	TSWAING METEORITE ~200KY
		NEOGENE	TERTIARY PERIOD	PLIOCENE	5.3	HOMINIDS APPEAR ~3MY
				MIOCENE	23	MAMMALS GREATLY DIVERSIFY DURING THE TERTIARY
		PALAEOGENE		OLIGOCENE	34	
				EOCENE	56	
				PALAEOCENE	66	
	MESOZOIC	CRETACEOUS			145	5th MASS EXTINCTION ~65MY MOROKWENG METEORITE IMPACT ~145MY
		JURASSIC			201	KAROO VOLCANISM ~183MY
		TRIASSIC			252	4th MASS EXTINCTION ~210MY FIRST MAMMALS DEVELOP
	PALAEOZOIC	PERMIAN			299	3rd MASS EXTINCTION ~252MY
		CARBONIFEROUS			359	FIRST REPTILES DEVELOP
		DEVONIAN			419	2nd MASS EXTINCTION ~354MY
		SILURIAN			443	FIRST LAND PLANTS DEVELOP
		ORDOVICIAN			485	1st MASS EXTINCTION ~440MY
		CAMBRIAN			541	CAMBRIAN EXPLOSION OF LIFE
PROTEROZOIC	NAMIBIAN*	1,000	PRECAMBRIAN	LATE PROTEROZOIC	1,000	EDIACARAN FAUNA ~560MY SNOWBALL EARTH ~700MY
	NAMAQUAN*	1,600		MIDDLE PROTEROZOIC	1,600	PILANESBERG COMPLEX ~1,200MY
	KHEISIAN*	2,057		EARLY PROTEROZOIC	2,500	OXYGENATION OF ATMOSPHERE ~1,900MY VREDEFORT IMPACT ~2,020MY BUSHVELD COMPLEX ~2,060MY
	VAALIAN*	2,650		LATE ARCHAEAN	2,800	VENTERSDORP VOLCANISM ~2700MY
ARCHAEAN	RANDIAN*	3,200		MIDDLE ARCHAEAN	3,200	1st MICRO-CONTINENT FORMS ~3,200MY
	SWAZIAN*	3,600		EARLY ARCHAEAN	3,600	1st BACTERIAL LIFE FORMS ~3,400MY EARLIEST OCEANIC CRUST ~3,500MY
				VERY EARLY ARCHAEAN	4,000	
			HADEAN		4,600	CREATION OF EARTH ~4,560MY

*The older South African-named Eras are now obsolete, and have been replaced by Epochs of the International timescale, as shown.

Adapted from the *International Chronostratigraphic Chart*

The Rock Cycle

It is useful to understand why and how there are continual changes within the rocks of the Earth's crust, albeit over geological time. The Earth is a dynamic planet and is in constant motion, a never-ending cycle of rock creation, modification, destruction and re-creation. The original theory is attributed to James Hutton, one of the early founders of modern geology, who in his famous 1788 treatise, *Theory of the Earth,* wrote: 'The result therefore of our present enquiry is that we find no vestige of a beginning and no prospect of an end.' This idea was popularised in Charles Lyell's *Principles of Geology* first published in 1830, which significantly influenced both Charles Darwin and road-builder and amateur geologist Andrew Geddes Bain, who lived and worked in South Africa. The essential point is that, over geological time, the rocks that make up the Earth are continually changing from one type to another by geological processes; put simply, the Earth is a great recycling machine. This is the concept of the Rock Cycle, as illustrated in **Figure 3** below.

The process starts with the creation of magma, which then cools and solidifies to form igneous rock. Exposed to the elements, this igneous rock becomes weathered and eroded, creating sediment, which is transported by wind and water and eventually deposited in a sedimentary basin, where it becomes compacted and cemented to form sedimentary rock formations. Under conditions of heat and pressure these sedimentary rocks may eventually become modified to produce metamorphic rock, which could become re-melted to produce magma, and thus the cycle is repeated.

This idealised cycle can be 'short-circuited': for example, under suitable conditions igneous rock can change directly into metamorphic

Figure 3: The fundamental Rock Cycle in which new rocks are constantly created as old ones are destroyed.

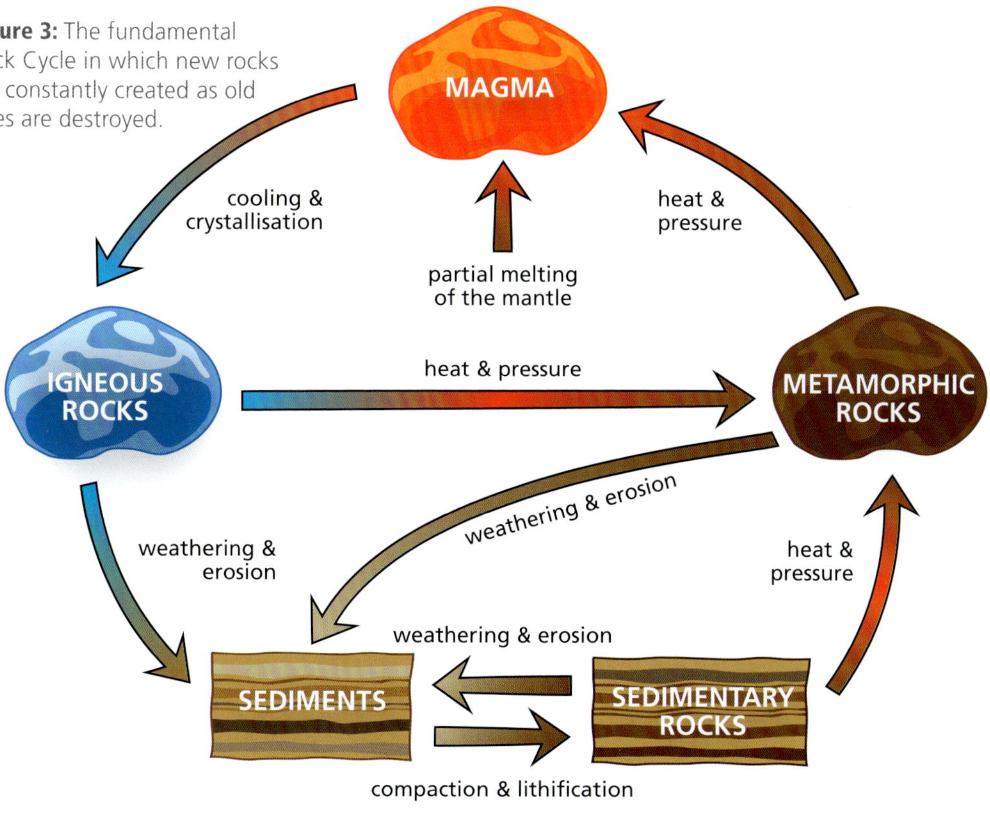

Plate tectonics – the movement of crustal plates

The Earth's rigid outer shell (the lithosphere) is composed of about 18 major plates and sub-plates that are able to move, albeit very slowly, on the underlying plastic upper mantle. These semi-rigid plates that make up the Earth's surface are constantly moving slowly relative to each other; they are said to 'jostle', an indication that Earth is a dynamic planet. Both oceanic crust (~7km thick) and continental crust (~35km thick) are extremely thin in comparison to the diameter of the Earth (~6,370km). It is at the boundaries or margins of plates that may be moving apart (divergent), or moving together (convergent) or moving past each other (transcurrent) that the dynamic interaction takes place, causing rifting, faulting, folding, thrusting, earthquakes and new igneous activity. Some of the overwhelming evidence includes:

- Sea-floor spreading (creating new oceanic crust) along the Mid-Atlantic Ridge where the African and South American Plates are moving apart at a rate of about 4cm per year (40km in a million years).
- The devastating earthquakes, tsunamis and regular volcanic activity around Indonesia, where the Indian-Australian and Eurasian Plates are converging and subduction is taking place (old crust being destroyed).
- Massive mountain building, deep crustal melting and volcanic activity in the Andes fold mountain chain along the west coast of South America, where the oceanic Nazca Plate is being forced under the continental South American Plate.

rock; or metamorphic rock can be weathered and eroded to produce sediment; or, as is often the case, old sedimentary rock can be recycled into new sedimentary rock.

The entire process is driven by the ongoing movement of crustal plates, known as the Theory of Plate Tectonics (see text box above). Plate tectonics explains the creation and closing of ocean basins, the thinning and thickening of the Earth's crust, and the creation of rising plumes of magma. The realisation that the Earth's crustal plates move relative to each other and that oceanic crust is much younger than continental crust was the key to understanding and explaining many fundamental geological phenomena.

The African plate has been almost stationary for the past 30 million years, and the surrounding plates are moving away from it. Thus Africa is the most stable of all the continents, although it is not entirely free of plate tectonic movement: the impressive and tectonically active African Rift Valley is a continental rift zone that extends the entire length of eastern Africa from the Red Sea to Mozambique. It displays a divergent plate boundary in the early stages of break-up, in which the Somali Subplate is in the process of splitting off from the rest of the African Plate. Centred in the Afar Depression, there is a classic triple-point meeting of the Gulf of Aden, the Red Sea and the Ethiopian Rift Valley, which continues southwest into East Africa. It is theorised to have been caused by a rising column of hot mantle material (also called a 'mantle plume') that formed deep beneath the crust, causing massive updoming. This resulted in the development of three rift arms at 120° to each other and considerable magma generation that caused the eruption of flood basalts. It is known as a mantle 'hot-spot'. There was once a similar but failed triple-point development in northeastern southern Africa that resulted in the formation of extensive Drakensberg and Lebombo lavas and Karoo dolerites.

Geological structures

Having looked at the broad picture, it's time to see what happens at a more local scale. Geology is not only about rocks and minerals but includes a variety of geological structures that affect the rock formations, sometimes in a major way. These include folds, faults, thrusts and joints, and landscapes and landforms are usually strongly controlled by these post-formational structures. Many geological exposures owe their spectacular appearance to preferential erosion along bedding layers, fault planes and joint surfaces.

Folds and folding

Folds are bends or undulations in sedimentary rocks that start off as being more or less horizontally layered. Folding is seen most clearly in layered, alternating sandstone-shale sequences, where the shaly beds assist with the differential movement of the stronger rock units that takes place during folding. Folding can result in conspicuous structural changes that happen at a wide range of scales, and it invariably involves changes in the nature of the original rocks.

Although often of random appearance, folding invariably follows a systematic and predictable pattern. Beds are initially folded into 'open folds' to produce anticlines and synclines, and with increasing tightness into steeply dipping isoclinal folds (**Figure 4**). Increased compression may result in the formation of overturned folds and eventually recumbent folds. Continued compression may result in low-angle thrust structures. In low-intensity, monoclinal folding only one fold limb is affected. Large-scale folding results in composite and complex fold structures called 'anticlinoria' and the complementary 'synclinoria'.

Folding results from horizontal compression exerted during a time of crustal 'closing' (e.g. when two crustal plates move together); the amount of folding varies considerably, depending on the degree of compression, depth of burial and the nature of the rock formations. When rock formations are deeply buried and strongly heated, folding is accompanied by a degree of recrystallisation, or even strong metamorphism. The Cape Fold Mountains of the Western and Eastern Cape show spectacular examples, and different styles of folding are evident in the same area, in some places the rocks apparently having experienced compression of very high intensity.

Figure 4: Diagrammatic geological fold structures produced by compression within the upper crust and, in places, by vertical movement within the basement.

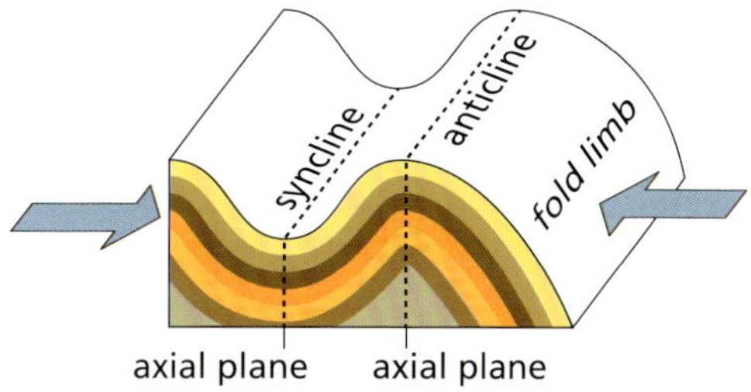

Open fold showing simple syncline and anticline

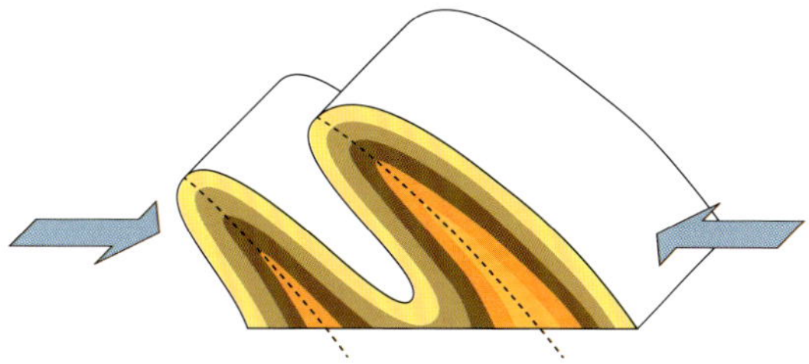

Overturned isoclinal fold – strata have similar dip

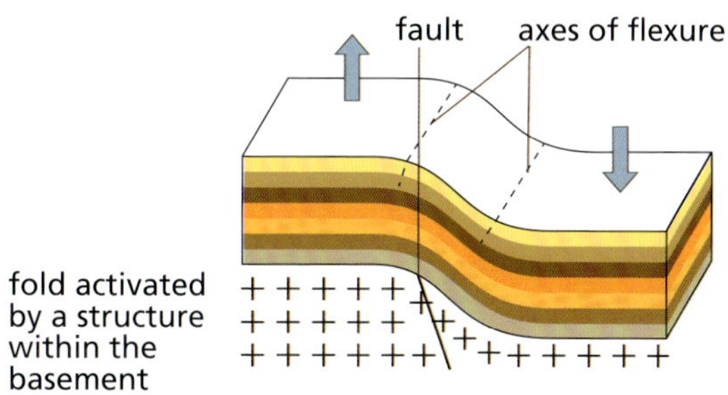

Monoclinal fold

Figure 5: Diagrammatic geological fault structures produced by different stress directions in the crust.

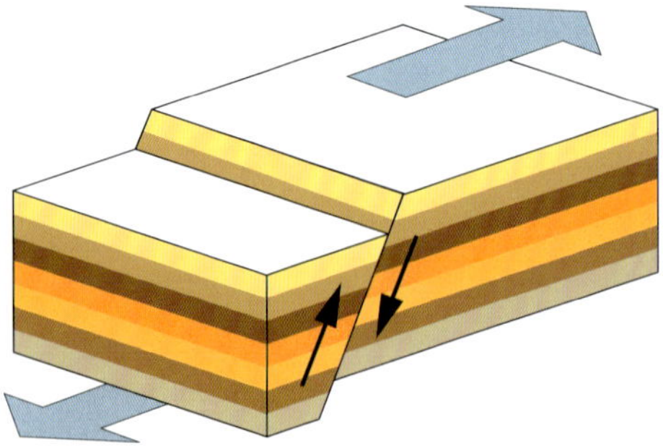

Normal or dip-slip fault

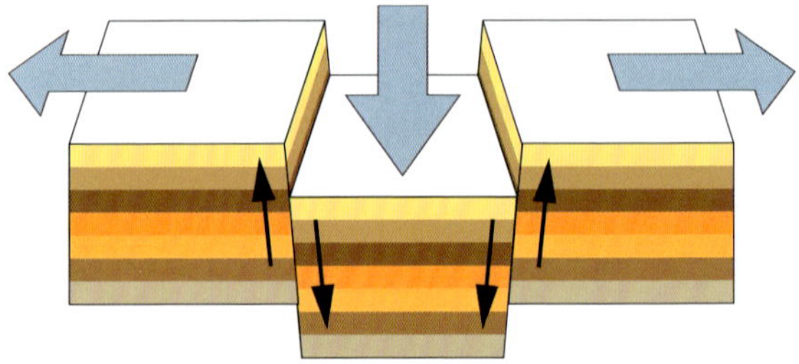

Parallel rift fault and graben structure

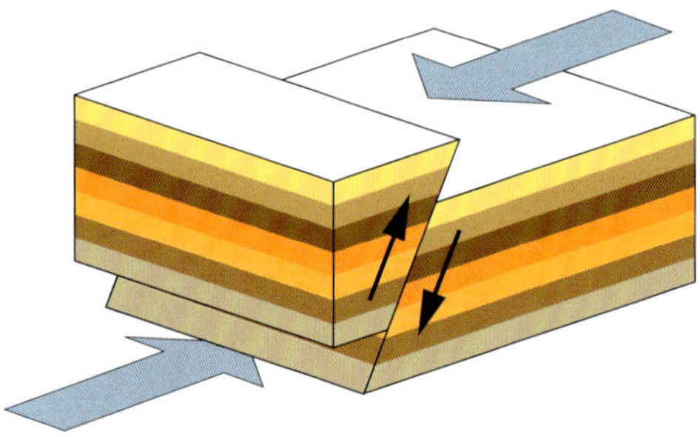

Reverse fault

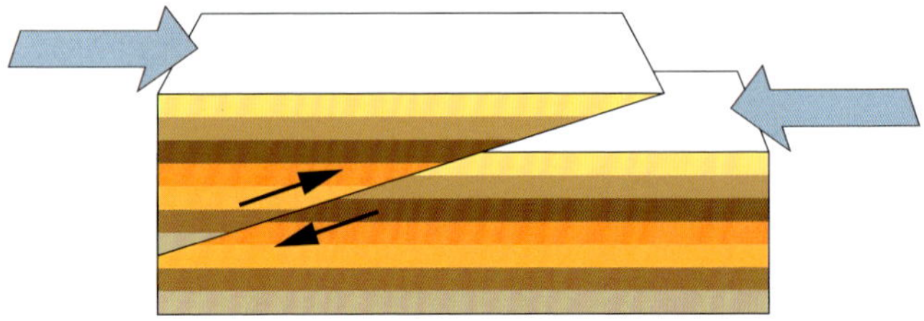

Thrust fault

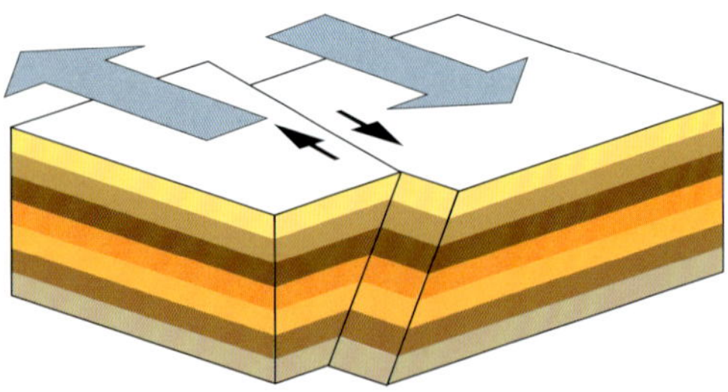

Wrench or strike-slip fault

Faults and thrusts

Faulting of most geological formations is very common and varies from small-scale to macro-scale, from mere centimetres to many kilometres in length, and of variable movement. A fault is generally a plane-like or slightly curved structure or discontinuity that forms where forces have caused a relative displacement of rocks on either side of the fault plane. If faulting occurs over a wide distance and comprises many smaller, individual faults, the combined effect is called a 'fault zone'. Depending on its location within the crust, the fault plane will show either brittle fracturing in the upper crust or ductile shearing in the lower crust. On a geological map, a fault will usually be seen as a line that represents where the fault plane projects to the surface. There are several types of faults, depending on the direction of displacement, and whether from tensional or compressive forces (**Figure 5**):

- **Normal or dip-slip faults** are most common, and are due to tension in the crust and include parallel **rift faults**. In South Africa, major normal faults include the Montagu and Worcester Faults, and the step-like faulting in the Soutpansberg seen at Wyllie's Poort (pp.214–218).

- **Reverse faults** result from compressive forces in the crust. Subduction zones form the largest of this type. An example of this is under the west coast of South America, where large-scale subduction has formed the Andes Mountain chain.
- **Wrench or strike-slip faults** are due to transform (sideways) movement. The San Andreas Fault in California is a major strike-slip fault zone.
- **Thrust faults** are similar to reverse faults but occur at a lower angle, when blocks slide over each other, as seen in Meiringspoort, for example (pp.86–89).
- **Ring faults** (not illustrated) can occur around collapsed calderas and may become filled by magma to form ring intrusions, as seen in the Pilanesberg (pp.248–253).

Fractures and joints

Joints in rocks are planes of brittle fracture that show preferred directions and regular spacing without movement. They are present in almost all surface outcrops and near surface rocks, and they play a major role in weathering and erosion. Almost any vertical rock surface or quarry face will reveal many joints that form a well-defined pattern of preferred joint directions. Joints generally occur as subparallel sets at more or less right angles, or as a strong parallel pattern. Close to the surface, joints are very susceptible to agents of weathering such as water, frost, wind and roots; they become the loci of erosion and play an important role in the development of landscape and landforms, notably controlling weathering in limestone that leads to the formation of caves. Regional scale jointing that is not easily recognised on the ground because we're too close to it often shows up well on Google Earth images. Several distinct types of joints exist and they include:

- **Tectonic joints** formed by subtle rock deformation resulting from minor regional stresses, probably due to earth movement.
- **Unloading joints** formed by release of the overlying compressive load due to erosion, allowing the rock mass to expand.
- **Exfoliation joints** that are similar to unloading joints, typically forming semi-horizontal sheeted layers in partly weathered granitic rocks.
- **Cooling joints** formed by contraction of hot rock masses, including polygonal columnar jointing in basalt or dolerite.

Cooling joints (or columnar jointing) formed within a thick, slowly cooling dolerite sill, seen here at the Valley of Desolation near Graaff-Reinet. Weathering has preferentially exploited the vertical joint planes between the solidified dolerite columns to produce this unusual but not uncommon dolerite landform (see pp.106–109).

Curved exfoliation joints are well developed in places on Paarl Rock and are common to many generally unbroken granite and gneissic bodies around South Africa (see Moon Rocks at Augrabies, p.299). This is the main mechanism for the peeling off of sheet-like granitic layers.

How landforms and landscapes are made

Landforms are natural physical features on the Earth's surface, while landscapes comprise the visible features of an area of land, in particular its natural scenery. The study of geomorphology focuses on the origin, evolution and processes involved in the making of landforms and the creation of landscapes, in which the role of geology is fundamental. Geomorphology is also of importance in disciplines such as agriculture, engineering, environmental studies, archaeology and even warfare.

The Earth's surface is in a continuous process of being worn down by forces of nature – part of the ceaseless Rock Cycle – although the process is almost undiscernible over a human lifetime. In simple terms, landscapes evolve over considerable time, starting with juvenile streams and steep valleys, moving to mature valleys and rounded terrain, and then to lowlands, senile rivers and flood plains. Eventually, this is followed by renewed crustal uplift – and the process starts again.

Geomorphological processes can be divided into four categories: the production of material by weathering; the removal of weathered detritus by erosion; the mass transport of the eroded material; and the deposition of the transported material. Weathering and erosion are geological processes that work closely together to shape the Earth's surface. The term denudation is often used to describe the lowering of a land surface.

Weathering

Rocks are slowly broken down into smaller mineral grains or to secondary minerals by chemical, physical or biological processes, often in combination; and the process is invariably accompanied by distinct changes in colour due to oxidation.

- **Chemical weathering:** This is the decomposition of some or all of the components by natural chemical reactions – a process of natural decay. Breakdown occurs by alteration of primary rock-forming minerals to secondary, softer minerals. For example, alkali feldspar, which decomposes

Deeply weathered and eroding Cape granite, seen on Chapman's Peak Drive near Hout Bay. Feldspar grains have been completely weathered to creamish clay minerals and the quartz is released as sand.

Weathering of the Msikaba Formation sandstone in the Oribi Gorge Nature Reserve is being assisted by a covering of lichen. This is a widespread, slow but potent mechanism for the biochemical breakdown of hard, quartz-rich rocks.

Along the Barberton Makhonjwa Geotrail, deeply weathered dolerite containing remnant boulders shows typical 'onion-skin' (small-scale exfoliation) weathering.

Weathered debris from the eroding high ground of Table Mountain has accumulated in lower areas to produce a soil-covered scree made largely of angular sandstone fragments, seen here in Diepsloot Reserve near Kloof Nek, Cape Town. This is largely a gravitational process.

to form clay minerals, particularly under wet, oxidising conditions. Very mildly acidic rainwater is especially effective in dissolving limestone and forming caves.

- **Mechanical weathering:** Such weathering involves the disintegration of rock by various physical means, including thermal expansion and contraction of minerals, freezing and thawing along fractures, abrasion by wind, grinding by glaciers, and attrition by wave and current action.
- **Biological weathering:** This describes actions such as those of lichens (see p.202) and mosses, of invasive roots causing expansion, of animals burrowing, digging and trampling, and of humans having an impact on the environment.

Erosion

The physical removal of decomposed or disintegrated rock material is very closely associated with weathering. Processes include transport of loose rock fragments of all sizes down mountain slopes (by gravity), along stream beds, in river courses, across plains and along coastlines, always to lower elevations. The main erosional agents are water, wind and ice.

- **Fluvial erosion:** Water run-off is by far the most important erosion agent, carrying away weathered rock debris by means of sheet-flow, and down gullies, streams and rivers. Eventually the transported material reaches a base level at a flood plain or sea, where the load is deposited. Major waterfalls are knickpoints created by changes in geological conditions. River terraces are evidence of slightly lowered base levels, and drowned valleys indicate major changes in sea level.
- **Aeolian erosion:** The sustained action of wind is an effective agent of erosion and sediment transport; it has given rise to today's sandy deserts and huge coastal sand dunes; and windblown deposits of the past were responsible for deposition of the extensive Clarens sandstone. Erosion of rock by wind is commonly seen in the Cederberg's 'sand-blasted' sandstones.

Glacial erosion: At times ice has been a significant agent of erosion and transport, having smoothed down land surfaces by glacial abrasion and plucking, and deposited the debris elsewhere. Ice-polished and striated rock surfaces occur at several places in South Africa, a relict of Dwyka glaciation around 300 million years ago.

Mass-wasting: In wet and steep conditions, soil and underlying weathered rock may move downhill under gravity, by means of soil creep, landslides, mud flows and rock avalanches. The process is both erosional and depositional. Relatively weak volcanic strata in the Drakensberg are prone to this process.

A juvenile mountain stream in the Drakensberg, filled with eroded basalt boulders derived from the surrounding mountains, seen here along Sani Pass.

At Coney Glen, near Knysna Heads, the action of waves along the southern Cape coast has produced a beach made of semi-rounded quartzite pebbles and boulders, a possible future conglomerate in the making.

The role of continental uplift on South Africa's landscape

Southern Africa's mostly elevated topography is unique, sitting as it does on a very stable continent, and untroubled by orogenic uplift for the last 1,000 million years, except along its southern margin in the Cape Fold Mountains. This is due to large-scale vertical uplift (or epeirogenesis) that has for many millions of years buoyed up the subcontinent, the cause of which is thought to be from anomalously hot conditions deep within the underlying mantle. This phenomenon is known as the African Super Swell, which is essentially a massive, long-lived mantle plume.

As far back as 180 million years ago southern Africa was already part of an elevated supercontinent. Following marginal rifting, the break-up of Gondwana and the creation of new coastlines, erosion on a vast scale resulted in the formation of new coastal plains that were separated from the high, slightly westward-sloping interior plateau by the rugged Great Escarpment that wraps right around southern Africa. Some 20–5 million years ago southern Africa again experienced periods of considerable continental uplift, up to 1,200m in the east, but much less in the west. These uplifts increased the elevation of the Great Eastern Escarpment, which led to large-scale marginal erosion and cut down deep valleys and gorges along the eastern and southern coastal plains. Of the once-widespread Drakensberg lavas, constant erosion has spared mostly the Lesotho Highlands, and exposed widespread 'fossil' land surfaces that predate the break-up of Gondwana.

South Africa's scenery – the big picture

We now look at the wide range of South Africa's spectacular landscapes, all of which have a strong geological basis; it helps to appreciate the big picture before moving on to the selected geological sites.

When you look at the landscape map of South Africa (**Figure 1**) you will notice that most of the country lies at a considerable elevation and slopes gently downwards to the west. In contrast to other continents, the vast interior plateau of southern Africa is around 1,000m–1,500m high and is separated from the much lower-lying and relatively narrow coastal strip by a very marked and irregular zone of rugged terrain, known as the Great Escarpment, a fundamental creator of many South African landscapes.

Over the years numerous 'geomorphic provinces' have been proposed. But in 1951 and again later, in 1967, Lester King famously described 18 such regions in his classic work *South African Scenery*, and defined them as 'regions of relatively uniform physiography that are more or less unique, although sometimes grading into one another'. Over time, other scientists have revised and added to the list so that now at least 34 geomorphic provinces have been determined, including sub-provinces, using modern methods.

On pp.34 and 35 is a summary of the major geomorphic regions of southern Africa, notably those relevant to the geological site descriptions in which the role of landforms and landscapes has been emphasised. More important regional landscape features are also described in the regional overviews and shown on the regional landscape maps.

A view from the Teekloof Pass that leads down the southern Great Escarpment from the Upper Karoo to the arid plains of the Great Karoo. The hills are made of horizontal Beaufort Group shale and lesser layers of harder sandstone, cut in the distance (top right) by two dipping dolerite sills that are part of a major system of Karoo intrusions in the area.

The major geomorphic provinces

- **Cape Fold Mountains:** A strongly linear region with two branches, the Northern branch fringed to the west by the Atlantic Coastal Plain, and the Southern branch bordered by the Southern Coastal Platform in the south; and it includes the Little Karoo.
- **Great Karoo:** An expansive semi-arid plain of usually dry watercourses and low koppies that lies inland from the Cape mountains and comprises the Roggeveld and the Tankwa and Southern Karoo areas. Because of underlying Karoo strata, some formations within the Main Karoo Basin stratigraphy are the main focus in the search for shale gas.
- **Great Escarpment:** A generally narrow, irregular and rugged strip completely surrounding the elevated interior plateau and separating it from the lower coastal margin of the subcontinent. It can be traced continuously from the Richtersveld and Namaqua Highlands in the northwest, around the Southern Escarpment, to the northern Drakensberg in the northeast. Its maximum development is where it meets precipitously with the Lesotho Highlands, in places higher than 3,000m.
- **Upper Karoo:** This vast plateau on the interior side of the Great Escarpment is underlain by Karoo strata and dolerite, and provides iconic 'Karoo koppie' scenery. It grades into the Bushmanland Plateau to the northwest, and into the Highveld plains to the northeast.
- **Southeastern Coastal Plain:** Lies to the southeast of the Great Escarpment and is crossed by numerous short rivers draining into the Indian Ocean. It is the coastal plain left by the retreat of the Great Escarpment, and now considerably modified. In the south it is shown by the narrow Southern Coastal Plain.
- **Eastern Lowveld:** This lies to the east of the Great Escapment in the northeast of the country and is mostly underlain by Archaean basement granite-gneiss. Still further east are the Lebombo Highlands, which comprise Karoo-age volcanic rocks.

Part of the hazy blue Soutpansberg range as seen from the north, along the N1. This elongated mountain massif was a barrier to road movement until construction of the road through Wyllie's Poort, seen entering the mountains in the far distance.

The Waterberg Plateau forms an impressive escarpment and mesas in western Limpopo, as shown here near Bokpoort Pass. It consists mainly of horizontal sandstone beds and is probably the Earth's oldest red bed sequence.

- **Highveld:** An extensive plateau of generally flat grassland on both sides of the upper Vaal River, extending east from the narrow Great Escarpment to the Kalahari in the west. To the north of this lies the Bushveld Basin.
- **Bushveld Basin:** A relatively lower area largely underlain by the Bushveld Complex and bounded to the north and south by stepped ridges. It includes the Springbok Flats and merges eastwards into the Mpumalanga Highlands.
- **Soutpansberg:** A major, fault-created mountain range in the north that formed within a rift-valley setting but now stands high above the surrounding plains of granite-gneiss of the Polokwane Plain and Limpopo Lowveld.
- **Limpopo Lowveld:** Lying north of the Soutpansberg, it forms the major part of the faulted Limpopo Valley. It is underlain by granite-gneiss and Karoo-age sediments.
- **Waterberg Plateau:** A large stepped plateau of flat-lying and thick sandstone that is bounded to the north and south by major fault zones. It is recognised as one of the world's first red bed sequences.
- **Orange and Vaal River Valleys:** The lower courses of these river systems have incised into the western part of the interior plateau; in the west the Orange River has cut deeply into the granites and gneisses of Namaqualand and Bushmanland.

The making of South Africa's landscape

Southern Africa's landscape is of considerable antiquity and parts of it date back to before the break-up of Gondwana. It has had a long and complex history, having passed through several major superimposed land-surface erosion cycles across southern Africa, as initially postulated by Lester King. The cycles are reflected by periods of large-scale denudation and planation (partial in places) that are correlated with times of broad regional uplift, and linked to equivalent periods of the deposition of sediment offshore. After uplift, the older existing, planed-down surfaces were re-activated and partly destroyed by a new period of erosion related to new base levels. Erosion surfaces throughout southern Africa can be recognised and, because of differing base levels, there are different erosion surfaces of the same age for the interior plateau and coastal areas.

The most important land-surface event was the formation of the long-lived African land surface after the break-up of Gondwana. It is estimated to have lasted 120 million years and formed a deep zone of weathering. It is uncertain what the land surface looked like before the break-up of Gondwana, but we do know that much of southern Africa was covered by Karoo flood basalt. King's ideas have been refined by others, notably by South African geoscientists Tim Partridge and Rodney Maud, whose simplified map of South Africa's erosion surfaces is shown in **Figure 6**. Relicts of the pre-African land surface are preserved in a number of high-lying areas, including the Cape Fold Belt, the Namaqualand Highlands, the Witwatersrand Highveld, the Soutpansberg and Waterberg and the Lesotho Highlands, and a number of geological sites are located in these ancient areas. The evolution of the southern African land surface since the start of the break-up of Gondwana is reasonably well established and is summarised in **Table 7**.

Figure 6: Simplified land erosion surfaces and major geomorphic regions of South Africa, dating from before the Gondwana break-up to Cenozoic times.

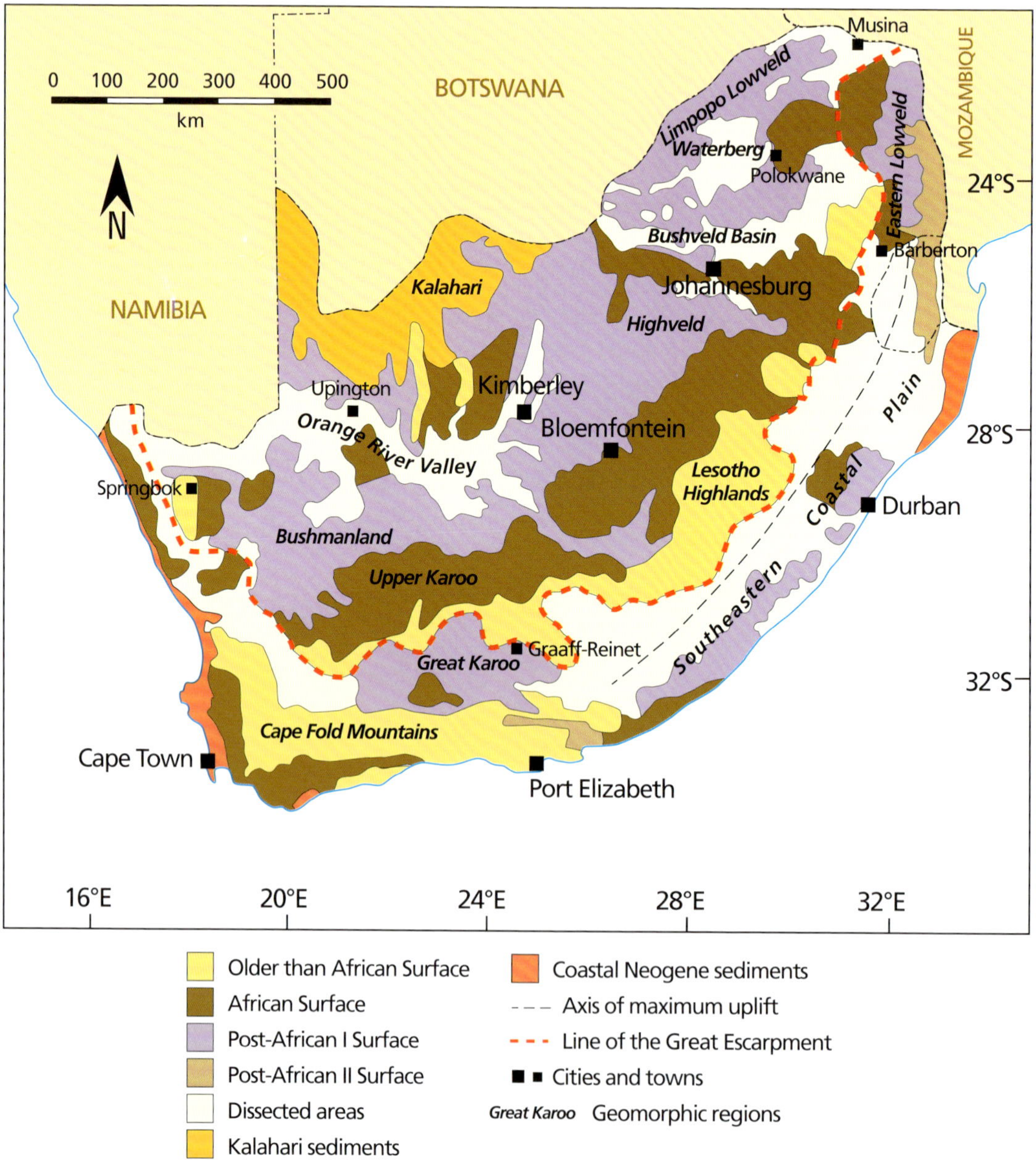

Adapted from Rodney Maud's chapter in *Southern African Geomorphology – Recent Trends and New Directions* (2012)

Table 7

LAND-SURFACE EVOLUTION IN SOUTHERN AFRICA FROM THE START OF GONDWANA RIFTING – FROM RECENT TO JURASSIC		
Major geological events	**Geological time period**	**Land-surface development**
Sea level changes due to climate fluctuations and glacial periods	<2.6 million years ago (Pleistocene to Holocene)	**Formation of marine benches,** coastal dunes, river terraces, coastal caves
Post-African II erosion cycle	<5 million years ago (Pliocene)	**Post-African II Surface** formed but of limited extent; renewed incision of rivers and coastal gorges
Major subcontinental uplift; up to 900m in the east, 200m in the south and 100m in the west	~5 million years ago (Pliocene)	**Termination of the Post-African I Surface;** westward tilting of the interior; seaward tilting of marginal erosion surfaces; new base levels set
Post-African I erosion cycle initiated; major axes of crustal warping renewed	~8 million years ago (Miocene to Pliocene)	**Post-African I Surface** formed (imperfectly); deposition of the Kalahari Basin
Moderate epeirogenic uplift; up to 300m in the east and 150m in the west; slight tilting westwards of the interior plateau	~18 million years ago (Miocene)	**Termination of African Surface;** new erosion base levels set in interior; the margin tilted towards the coast
African erosion cycle, a multi-cyclical event; extends through the long break-up of Gondwana	~140 to ~20 million years ago (Jurassic to Miocene)	**African Surface** formed over a long period and at different levels; major erosion, extreme planation and deep weathering
Gondwana break-up starts and continues for more than 30 million years; strong rifting and formation of new coastline	~150 to ~120 million years ago (Jurassic to Cretaceous)	**Proto-Great Escarpment** starts along new coastline; new base levels of erosion are set in the interior and along the coast
Mantle hot-spot developed under Gondwana; rift faulting initiated and Karoo magmatism starts	~180 to ~175 million years ago (Jurassic)	**~1,500m of Drakensberg basalt lavas** flood the continent as precursor to the break-up of Gondwana; land elevated mostly in the east

Adapted from Rodney Maud's chapter in *Southern African Geomorphology – Recent Trends and New Directions* (2012)

WESTERN CAPE

OVERVIEW OF LANDSCAPES AND GEOLOGY

This chapter focuses on six geological sites in the Western Cape (Geosites 1–6) which are shown on the national maps (pp.12–13) as well as on the regional landscape and geological maps (pp.40–41). Geological formations relevant to the selected geosites range from metamorphic rocks of the Late Proterozoic to young coastal deposits of the Tertiary, as outlined below. Notably, the geology of the impressive Cape Fold Mountains continues eastwards into the southern and Eastern Cape.

The major geological formations of the region are listed below:

Malmesbury Group
(Late Proterozoic)

These are the oldest rocks in the region, formed around 600 million years ago. They exist as inliers of strongly deformed and metamorphosed sedimentary formations, created during the making of the Gondwana supercontinent. They are mostly dark laminated shales, siltstones and some sandstones, and they reveal no recognisable fossils. Easily weathered and usually poorly exposed, they have not had a strong influence on the landscape. However, excellent exposures occur along the Atlantic seaboard, north of Sea Point (**Geosite 1**).

Cape Granites (mainly Cambrian)

During the creation of the Gondwana supercontinent, a variety of large granitic bodies were intruded at depth into the Malmesbury Group, 555–515 million years ago. They occur mostly from Vredenburg to Stellenbosch and show a wide range of mineral composition and texture. After considerable crustal uplift, followed by large-scale denudation, the Table Mountain sandstones were deposited in shallow water on a planed-down granitic or metamorphic basement. Today, they generally form rolling, sandy country but, in places, rapid erosion has produced spectacular exposed domes such as seen on the lower slopes of Lion's Head and Paarl Mountain. Places to note are south of the Sea Point Contact (Table Mountain, **Geosite1**); at Chapman's Peak Drive (**Geosite 2**) and Paarl Rock (**Geosite 3**).

Cape Supergroup
(Ordovician to Carboniferous)

This originally 8,000m-thick sequence was deposited 500–330 million years ago in an elongated sedimentary basin, part of the ancient Agulhas Sea that skirted the southern part of Gondwana. It is continuous and extends for more than 1,000km across the southern part of South Africa, from Nieuwoudtville in the northwest, and along the southern Cape to north of Port Elizabeth. The full sequence comprises three major geological groupings known, stratigraphically from oldest to youngest, as the **Table Mountain**, **Bokkeveld** and **Witteberg Groups**. Made almost entirely of hard quartzitic sandstone, it is the 4,000m-thick Table Mountain Group that generally dominates, while intermontane valleys are usually underlain by less resistant Bokkeveld rocks, or filled by much younger formations. The sandstone comprises material eroded off the highlands of Gondwana and deposited on coastal flood plains or in the shallow sea. In its lowermost part the group consists of alternating reddish shaly rocks and sandstones, as seen at Chapman's Peak Drive (**Geosite 2**), followed by a 2,000m-thick sequence of clean, hard sandstone, seen unfolded on the cliffs of Table Mountain (**Geosite 1**). Stratigraphically above this is a further thick sequence of more brownish-weathering, well-bedded sandstones that are best seen in the Cederberg (Stadsaal Caves, **Geosite 5**). The full Table Mountain sequence is present on the road through Cogmans Kloof (**Geosite 6**), dramatically folded in places.

CAPE MOUNTAIN BUILDING

(Gondwana Orogeny, Permian to Triassic)

Many millions of years after the Cape Supergroup had become compressed into layers of hard rock, a zone of crustal subduction developed along the southern margin of Gondwana. As oceanic crust was thrust under continental Gondwana, strong compression of the sedimentary formations took place, which buckled and eventually deformed into enormous, complex folds, accompanied by low-angle thrusting. This took place episodically 280–230 million years ago and the outcome was a linear fold mountain range of Himalayan proportions that extended across the entire supercontinent. Today this exists as the worn-down but still impressive Cape Fold Mountains that form both inland and coastal ranges, and extending from the Cederberg in the west to the Zuurberg in the east. While the enormous mountain-building event was taking place, the continental crust to the north of the mountains began to sag and the new Karoo Basin developed, allowing the sea to enter.

A simplified interpretation of the very complexly folded zone in the mountains at the northeast end of the kloof, close to Montagu. These once horizontal beds of Nardouw sandstone in the upper part of the Table Mountain Group have been intensely compressed and deformed, possibly due to gravitational collapse. The white dashed lines indicate fold structures. Very similar folding is seen in the Great Swartberg range.

Sandveld Group

(Tertiary to Quaternary)

This is one of the youngest groups in southern Africa, and is restricted to the slightly elevated coastal plain of the Western Cape, from False Bay in the south to Elands Bay in the north. Good outcrop is largely obscured by even younger dune sands and soils. In the area east of Saldanha Bay several poorly consolidated formations were deposited on an uneven basement, such as Cape granite at Langebaan. These young rocks of variable origin formed during a series of major marine transgressions and regressions during the Tertiary, and the area is well known for a wide variety of fossils of Miocene-Pliocene age. These can be seen in the phosphate-rich sediment exposed in the old Varswater phosphate quarry near Langebaan Road, the location of the West Coast Fossil Park (**Geosite 4**).

Figure 7: Landscape map of the Western Cape region. The higher ground of the Western Cape, coloured in tones of red and yellow, is dominated by the presence of the Cape Supergroup, as clearly seen when comparing this map with the geological map alongside it. Featured geological sites appear in red circles.

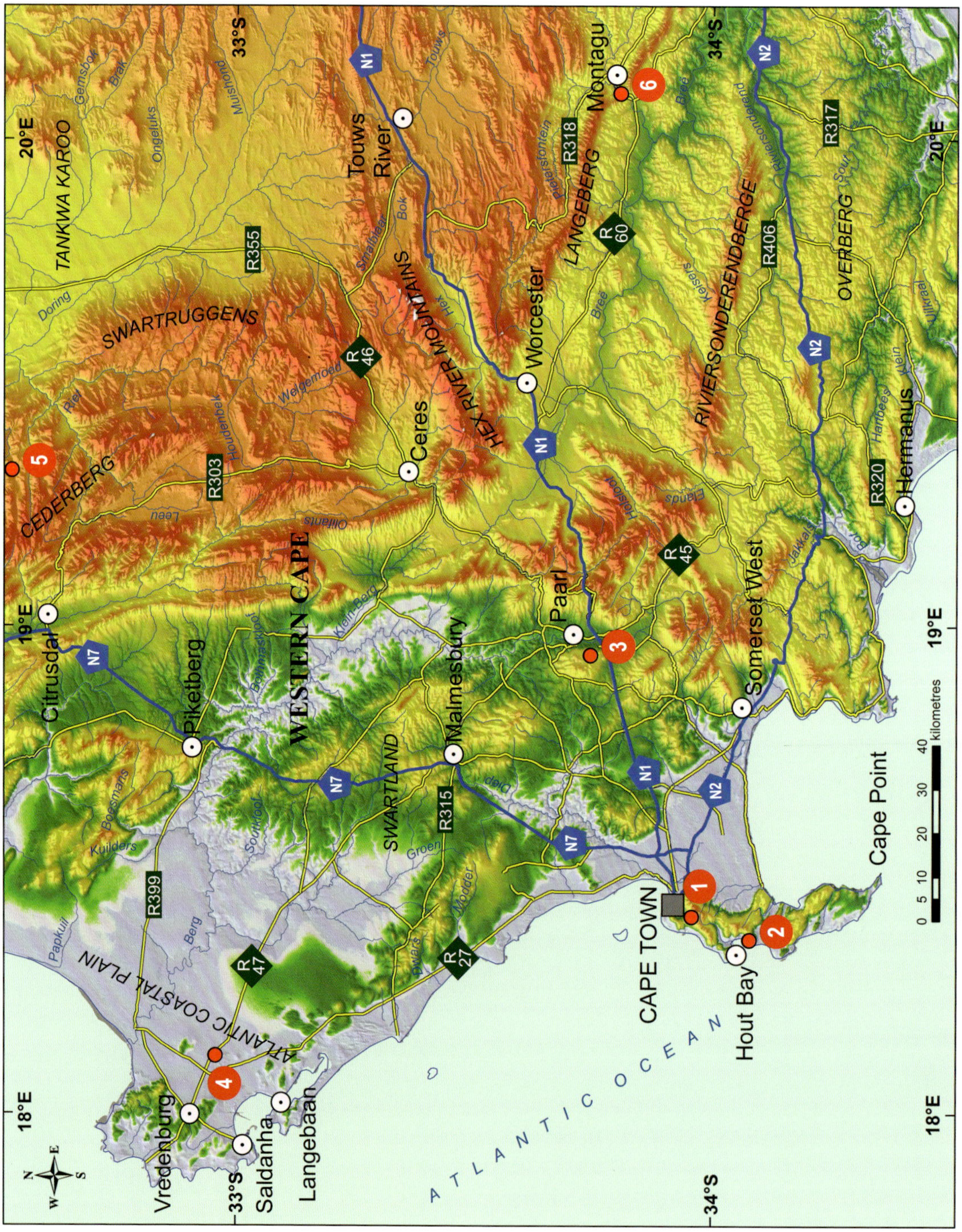

Figure 8: Geological map of the Western Cape region. The Western Cape is dominated by the Cape Supergroup, shown in shades of blue, which lies on an older basement of deformed Malmesbury Group rocks, shown in light brown and on Cape granite, in pink. The start of the Karoo Basin is seen in the northeast.

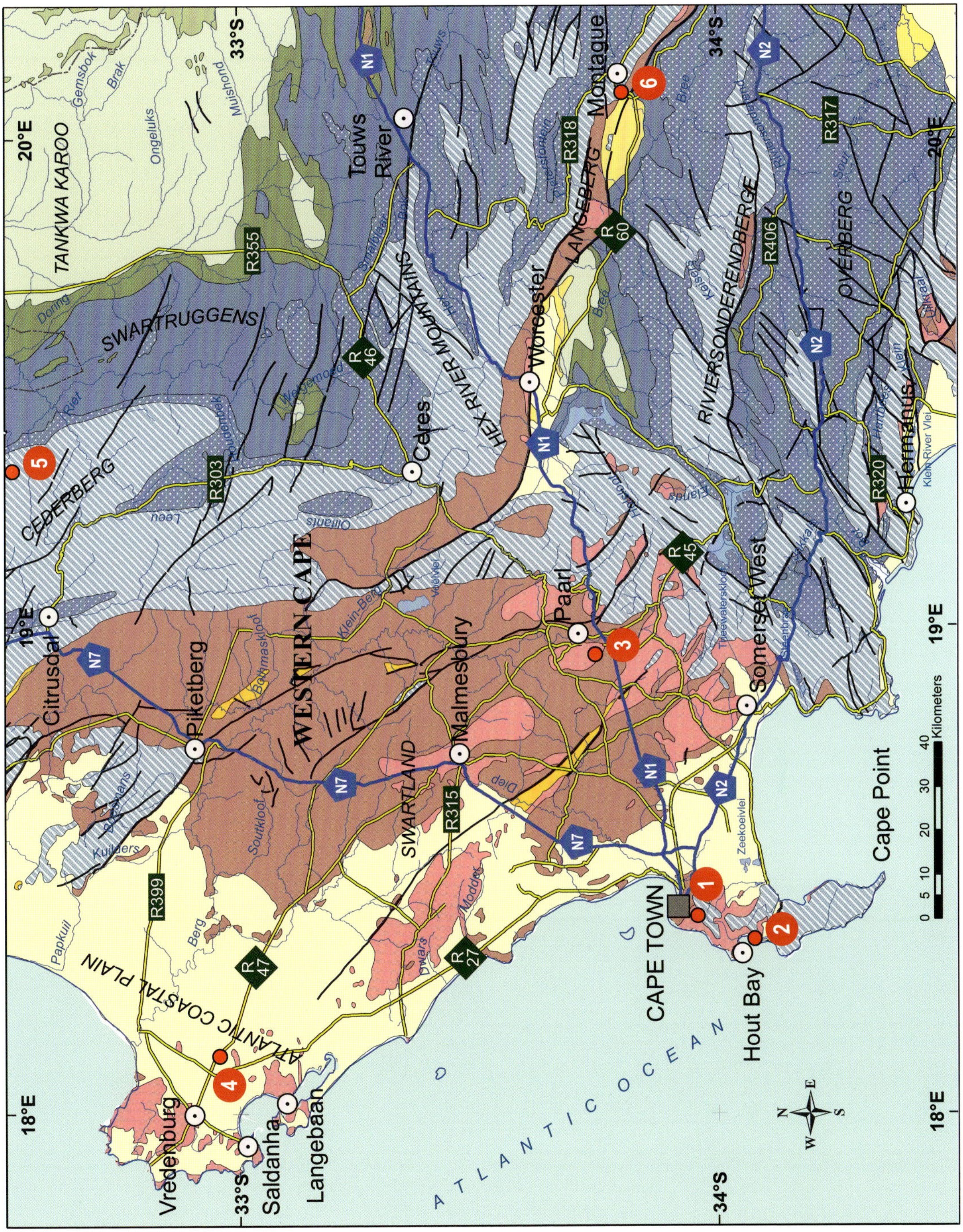

• A stratigraphic key to this simplified regional geological map appears on the inside covers

1

Table Mountain

WORLD HERITAGE SITE

One of the world's 'Seven New Wonders of Nature'

Shaen Adey / Images of Africa

Devil's Peak, on the left, and Table Mountain, as seen from Signal Hill. The cliff-like, layered upper formation is hard sandstone of the Peninsula Formation, which overlies the reddish sandstone and siltstone beds of the Graafwater Formation. The position of Tafelberg Road mostly follows the base of the Graafwater sedimentary rocks, which overlie poorly outcropping Malmesbury Formation rocks or Cape granite.

Iconic Table Mountain is a geological landmark known all over the world, and there is no sight more evocative of South Africa than this unique, amphitheatre-like mountain, with Cape Town nestling at its foot. In reality, the bold north face is part of a mountain chain that extends to Cape Point, some 50km to the south.

Johan Dempers www.aerialimage.co.za

■ Location and getting there

- Finding Table Mountain is never a problem and on a good day it is visible from at least 100km out to sea. The easiest way to the top is via the famous aerial cableway, starting from the Lower Station on contour-hugging Tafelberg Road. There are also numerous hiking routes up the mountain from different sides.

- GPS (Lower Cableway Station): 33°56′38″S; 18°24′11″E

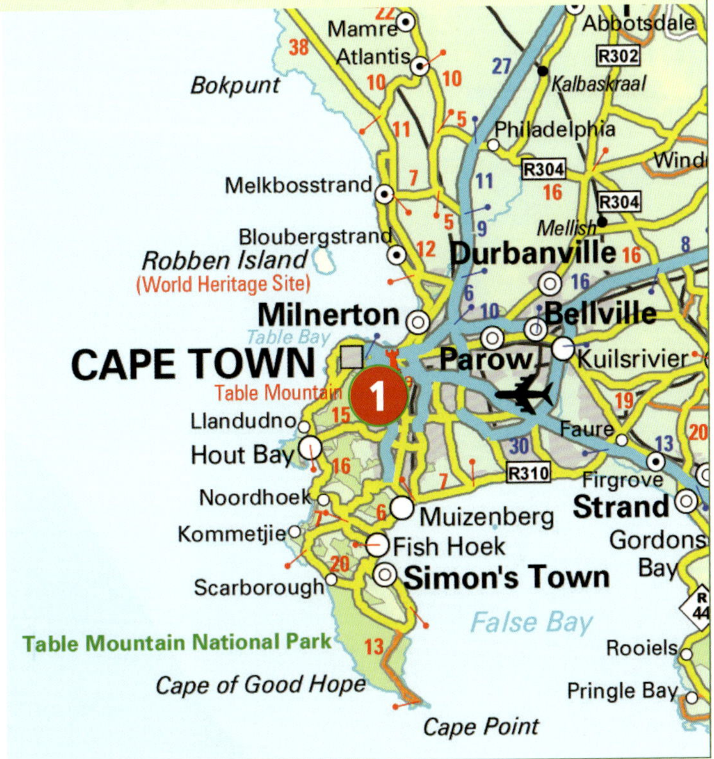

Setting the scene

Although the mountain was probably climbed by Portuguese maritime explorer Bartholomew Dias in 1488, the first record of its ascent was by Antonio de Saldana in 1503. Long before that it was called Hoerikwaggo, or Mountain at the Sea, by the Khoisan. Dutch explorers named it Tafelberg and in 1652 a replenishment stop was established in Table Bay on the sea route to the East. The mountain falls within Table Mountain National Park and is a World Heritage Site, one of the Cape Floral Region's Protected Areas.

Opposite: A south-looking aerial view of Table Mountain showing several important landmarks: the igneous contact between Malmesbury metamorphic rocks and intrusive Cape granite is shown by the dashed white line starting at the Sea Point Contact and ending on Tafelberg Road close to Devil's Peak, where it is overlain by younger Table Mountain sediments.

Regional geology

The rugged heights of the mountain are made of horizontal strata of the **Table Mountain Group**, part of the **Cape Supergroup**. The basement consists of either steeply dipping metamorphosed sedimentary rocks of the **Malmesbury Group**, or the **Sea Point granite**, which is part of the 540-million-year-old **Cape Peninsula Batholith**, an intrusion of the **Cape Granite Suite** (an idealised north-south cross-section of Table Mountain is shown in **Figure 9**).

Landscape and landforms

Table Mountain is a residual mountain that literally rises out of the sea, and it has a long and complex geological history, having been involved in major cycles of deposition, mountain building and granite intrusion. During a time of much higher sea level 5 million years ago, Table Mountain and the rest of the peninsula formed two offshore islands, when the sandy Cape Flats and Noordhoek Valley were covered by the sea. The remarkably flat Table Top is about 2.5km wide, reaches a maximum elevation of 1,086m at Maclear's Beacon and slopes gently to the south. To the northeast it is flanked by Devil's Peak (1,000m), while the butte-like peak of Lion's Head (669m) lies to the northwest. At one time the sandstone formations would have been continuous between these features. The sandstone is well jointed and is being actively eroded by rock falls, notably along faults. When compared to other mountains of the Cape Fold Belt, the flat mountain top is an enigma, and is attributable to the underlying rigid granite batholith that resisted deformation, and to the hard, horizontal layers of sandstone that protect underlying formations from erosion. There is also the remote possibility that Table Mountain is a remnant of a mega-scale synclinal fold, now almost completely eroded.

South African Tourism

Massive vertical walls of hard, blocky sandstone in the rugged ravine leading to the Upper Cableway. Notice the cable-cars, which slowly rotate to give panoramic views.

The top of Table Mountain is relatively flat due to the almost horizontal bedding, although there are many eroded sandstone outcrops near the escarpment edge.

The Upper Cableway Station (at 1,067m) on the northwestern edge of the Table Top, with Lion's Head (669m) in the distance.

Horizontal layers of hard, well-jointed, blocky Peninsula Formation sandstone can be seen from the rotating cable-car. Erosion is largely by mechanical breakdown of jointed sandstone blocks.

Rock formations

- The towering cliffs and buttresses are made of some 500m of well-bedded, light grey, quartz-rich Peninsula Formation sandstone, dipping at about 5° to the south. The sandstone contains scattered quartz pebbles and some thin layers of conglomerate, indicating deposition in shallow water.
- Underlying this is a 70m-thick unit of brownish Graafwater Formation comprising alternating 'mixed' sedimentary layers, exposed beautifully along Tafelberg Road (also see Chapman's Peak Drive, pp.49–53). Preserved in isolation at Maclear's Beacon is a remnant of the overlying Pakhuis Formation tillite of glacial origin.
- Dark, deformed and metamorphosed shaley rocks of the Malmesbury Group are also seen in places along Tafelberg Road, and Signal Hill is entirely underlain by this poorly outcropping formation.
- Large dome-shaped outcrops of Sea Point granite occur on the seaward slope of Lion's Head but at Kloof Nek the same granite is generally strongly weathered and rotten, forming a light-coloured, clay-rich cover.

Close-up of Peninsula Formation sandstone outcrop on top of the mountain. On site you will see the gritty layers and the slight dip of the strata to the south.

Table Mountain is being actively eroded away. This white sand, now on the mountain top and derived from weathered sandstone will, sooner or later, end up on a local beach.

The western edge of Table Mountain seen from Kloof Nek Road. The strongly dissected Twelve Apostles to the south result from a series of northwest-trending faults that have formed fracture zones between the buttresses and assisted mechanical breakdown. The approximate location of the sedimentary unconformity between underlying older Sea Point (Cape) granite and Table Mountain Group sandstone is shown by the dashed white line.

Platteklip Gorge cuts through the front face of the mountain and provides a popular steep zigzag path from Tafelberg Road to the top, with excellent geology all the way. It has developed along a zone of strong fracturing.

The sandstone butte of Lion's Head from Sea Point, showing the underlying weathered granite dome, similar to that seen at Paarl Rock.

Finely cross-bedded and planar-bedded sandstone, indicating deposition in shallow water, in the Graafwater Formation along Tafelberg Road.

Coarse-grained granite is deeply weathered to clay in places, here seen near Kloof Nek Road, located in the topographic saddle that separates Table Mountain from Lion's Head.

Figure 9: An idealised north-south cross-section through the Table Mountain massif shows the main geological elements, geomorphic features and other landmarks.

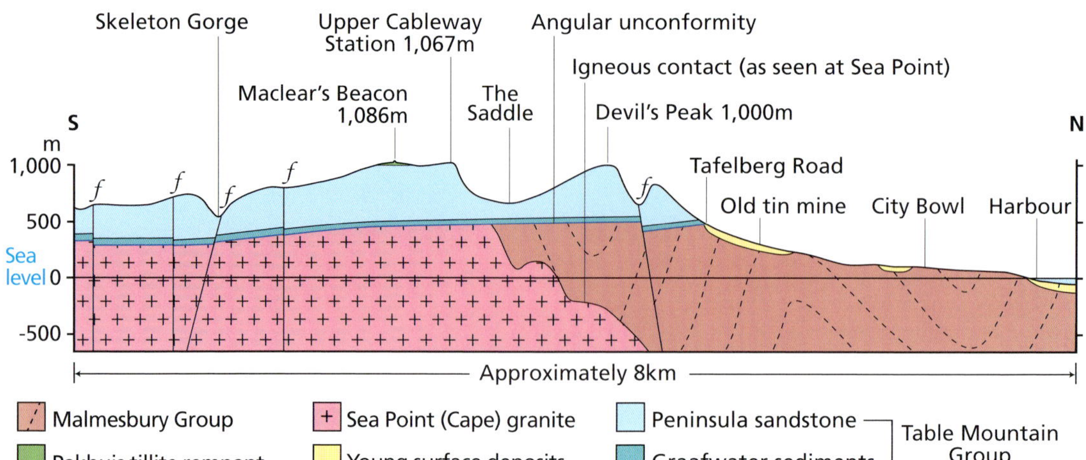

Adapted from Council for Geoscience geological map series (1984)

Typical alternating sequence of the 70m-thick Graafwater Formation, consisting of well-bedded sandstone, siltstone and maroon shale, seen along Tafelberg Road.

Geological history

- After deposition of the Malmesbury Group, the shales were deformed and metamorphosed during the Saldanian Orogeny around 560 million years ago, and intruded at depth by Sea Point (Cape) granite magma.
- Over many millions of years the uplifted mountains were eroded and planed down to sea level, as seen from the unconformity on Chapman's Peak Drive.
- About 450 million years ago the new Agulhas Sea created a depositional basin more than 1,000km long, which slowly filled with sediments derived from the Gondwana hinterland, and formed the Cape Supergroup, deposited in two significant stages:
1) At first, sediment was deposited on a broad coastal flood plain crossed by meandering rivers, depositing the alternating layers of sand, silt and mud that make up the approximately 70m-thick Graafwater Formation.
2) Following this, vast amounts of clean quartz sand settled on top to form the thick sandstone beds of the Peninsula Formation. A later glacial period is reflected by the remnant of Pakhuis tillite at Maclear's Beacon.
- Later large-scale folding of the Cape Fold Belt during the Gondwana Orogeny (see Western Cape Overview, p.39) left Table Mountain largely unaffected.
- Following the break-up of Gondwana, considerable uplift and erosion took place, and deeply buried Malmesbury rocks and Sea Point granite were exposed, resulting in today's Table Mountain massif.

The famous Sea Point Contact

This historic site and National Monument (now a Provincial Heritage Site) is one of the world's classic geological outcrops, made famous by Charles Darwin following his visit to Cape Town in 1836 on HMS Beagle, and it was influential in the early understanding of Earth processes. Excellently exposed along the rocky Atlantic shoreline is the igneous contact between light-coloured, coarsely porphyritic Sea Point (Cape) granite and blackish, thermally metamorphosed shales (now hornfels) of the Malmesbury Group. The contact zone is complex, extends over more than 150m, and reflects a time around 540 million years ago when molten granite magma intruded into folded Malmesbury sedimentary rocks, then at considerable depth below the surface. The coarse-grained granite intrusion into the shale formed a zone of mixed rock.

Part of the Sea Point Contact showing coarse-grained granite (left) and dark metamorphosed shale (now hornfels) to the right. Molten granite magma intruded the already deformed shale to produce a mixed 'migmatite' zone at the contact. The granite also contains small dark inclusions (xenoliths) of hornfels.

Things to see and do

- **The cableway:** An effort-free way to the top of Table Mountain takes you less than 5 minutes and provides exceptional views. See the hard, jointed sandstone cliffs on the way up and the weathered sandstone geology on top.
- **Walks on the Table Top:** There are several interesting walks, the longest being to the highest point at Maclear's Beacon (1,086m) on the east side, where you can see a remnant of Pakhuis tillite.
- **Hikes up the mountain:** There are several strenuous hiking routes, the most popular being the zigzag path up Platteklip Gorge, formed in a zone of strong fracturing, leading off from Tafelberg Road.
- **The Lion's Head hike:** Try the not-so-arduous walk up this isolated peak, a remnant of the once continuous layer of Table Mountain sandstone.
- **The old tin mine:** From 1910–1912 mining took place close to suburban Vredehoek, on the lower slope of Devil's Peak. Look out for the 55m-deep vertical shaft and a 120m adit that both still exist, and well as other ruins.

2 Chapman's Peak Drive

WORLD HERITAGE SITE

A spectacular sedimentary unconformity along one of the Cape's most scenic drives

Fiona J Hinds

An aerial view of Chapman's Peak and its remarkable 'drive' seen from the west, in the late afternoon sunshine. Notice the road that follows the gently southward-dipping sedimentary unconformity between Cape granite below and the thick sequence of Table Mountain sandstone above. In the distance on the right is the Noordhoek Valley.

Chapman's Peak is an imposing 593m-high mountain on the Atlantic side of the Cape Peninsula. A spectacular coastal pass, known as Chapman's Peak Drive, hugs its precipitous western side. The road has recently been re-engineered to protect motorists from dangerous rock falls.

Location and getting there

- Chapman's Peak Drive is part of Cape Town's M6 route that connects Hout Bay to Noordhoek, and to Cape Point beyond. It lies within the Table Mountain National Park and is part of the Cape Floral Region's World Heritage Site. The spectacular drive (now a toll road) starts south of Hout Bay and winds along the rocky coastline for about 7km before heading inland at Noordhoek.

- GPS (Hout Bay viewing site): 34°04′39″S; 18°21′25″E

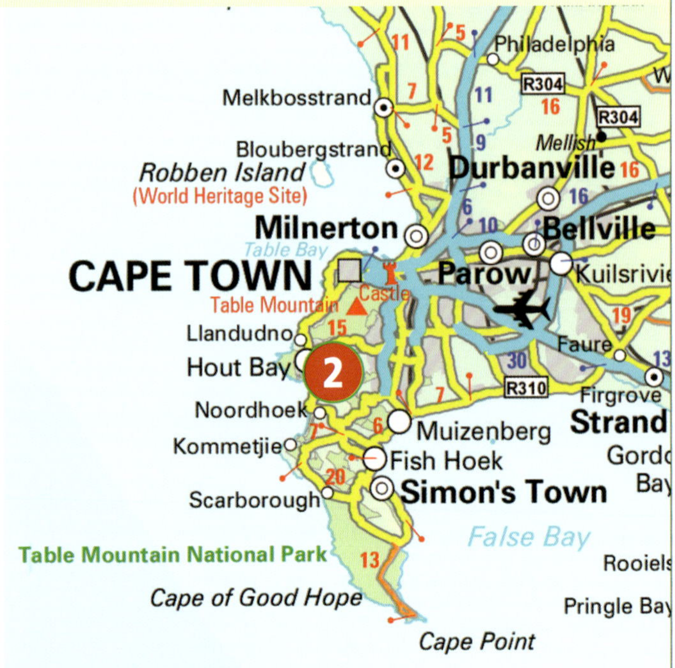

Viewed south from near the north-end viewpoint, this shows the road as it follows mostly the base of the well-bedded, reddish-brown Graafwater Formation, which lies unconformably on massive, partly weathered Cape granite.

Setting the scene

Nearly 600m high, Chapman's Peak drops precipitously into the sea over a horizontal distance of less than 500m. The magnificent coastal pass, regarded as a major feat of design and technology at the time of its construction between 1915 and 1922, was cleverly engineered to follow precisely the geological contact between hard, planed-down, older Cape granite and the younger, more workable overlying Table Mountain sedimentary formations. Road cuts reveal in great detail the geology at the contact and of the overlying sedimentary rocks. Regrettably the necessary recent re-engineering has covered some of the fine geology.

Regional geology

There are two very different formations: **Sea Point granite**, part of the 540-million-year-old **Cape Peninsula Batholith**, an intrusive of the **Cape Granite Suite**; and overlying sedimentary rocks of the **Table Mountain Group** that form the base of the **Cape Supergroup**. They are separated by the celebrated unconformity (or nonconformity) that reflects a time gap of many millions of years – one of the finest examples in the world of sedimentary rocks overlying granite. This contact slopes gently to the south and shows minor undulations on the older, planed-down granite land surface.

Landscape and landforms

Elevated and partly dissected peaks of the Table Mountain system make up the landscape, actively being worn down by erosion. Below road level the granite outcrops as bare rounded boulders, fresh because of constant erosion close to the sea; above the road the sedimentary rocks form impressive ramparts and cliffs, and are an ongoing source of rock falls – an unstoppable natural process.

This area, too, was part of an offshore island about 5 million years ago (see Table Mountain, pp.42–48). The steep mountain slope and cliffs have been preserved by the hard layers of erosion-resisting horizontal sandstone.

Rock formations

- The Sea Point granite is light grey, coarse-grained and strongly porphyritic (of varying grain size with large rectangular crystals of alkali feldspar, called 'phenocrysts'), and is well exposed along the coast between Sea Point and Hout Bay.

Top: The Graafwater Formation comprises an alternating sequence of very well-bedded, buff-coloured sandstone, finer siltstone and maroon shale, deposited under changing conditions and sometimes exposed to the air, and so imparting the oxidised maroon colour. Cape granite at the bottom of the road cutting is not seen in the photo. Inset: A detail of the Graafwater sedimentary rocks. Here, at the time of deposition, a bed of maroon mud (now shale) was disrupted by small slumped masses of more compacted sand (now sandstone) to form pillow-like bodies.

- The overlying Graafwater Formation consists of a 70m-thick alternating sequence of thinly bedded, reddish-brown sedimentary rocks that indicate a changing environment at the time of deposition. These comprise buff-coloured sandstones, siltstones and maroon mudstones, often showing sedimentary depositional structures.
- Above this are the much thicker and more resistant, light grey, cliff-forming sandstones of the Peninsula Formation. There are narrow cross-cutting dolerite dykes and minor faults.

Geological history

- After the granite was intruded (during the Saldanian mountain building, approximately 540 million years ago), there followed a long period of erosion and planation that resulted in the creation of a vast coastal plain adjacent to the ancient Agulhas Sea.
- Around 450 million years ago meandering and braided rivers draining the Gondwana highlands deposited layers of sand, silt and mud on both the coastal flood plain and also under shallow tidal conditions to form the Graafwater rocks. Exposure to the atmosphere resulted in mudstones having a maroon colour as a result of oxidation.
- The coastal plain was subjected to periodic flooding by the sea and during these times deposition became marine. Large quantities of clean quartz sand were deposited in a shallow sea, atop the Graafwater layers, to become the thick Peninsula sandstone.
- Deep burial of the Cape Supergroup took place, followed by the period of Cape mountain building, although the formations seen on the Cape Peninsula were not deformed in the process.
- Later renewed uplift after the break-up of Gondwana and strong coastal erosion resulted in the landscape seen today.

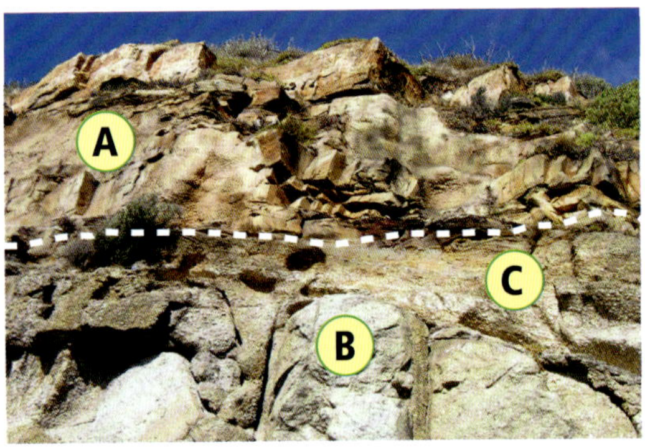

*This exposure reveals the unconformity (marked with a dashed white line) between horizontal Table Mountain sandstone (**A**) lying on top of older Cape granite (**B**). The buff-coloured zone seen at the top of the granite (**C**) is a reminder of an earlier weathered land surface that existed.*

Close-up of partly weathered, strongly porphyritic Sea Point granite. The blocky feldspar crystals (phenocrysts) are more than 5cm in length.

Massive fractured granite outcrops occur below the road (off-photo to the top right), dropping away steeply into the sea. In the background is Hout Bay and the mountain called The Sentinel.

The historic manganese mines

Long-abandoned manganese mines, found on the higher ground at the Hout Bay end of Chapman's Peak Drive near East Fort, are a geological curiosity. Around 1910, manganese ore, mainly pyrolusite (MnO_2), was mined in some quantity from veins and vertical lodes formed in Table Mountain sandstone. Adits were blasted into the mountainside and, for a time, rich ore thundered down a 750m-long iron chute directly to a jetty built on the rocky coast, and was then exported to England. The ore soon pinched out, but the endeavour is still remembered by a display at the Hout Bay Museum.

John Rogers

One of the deep, mined-out vertical fissures at the old manganese mines that are found in the mountain near Hout Bay.

Chapman's Peak Drive at the Noordhoek end, where you can see one of South Africa's longest beaches, with Kommetjie and the Slangkop lighthouse in the distance.

Deeply weathered and easily eroded granite along the Hout Bay side of Chapman's Peak Drive, overlain by Table Mountain sedimentary rocks.

Things to see and do

- **The toll road:** The coastal mountain scenery along this drive is unsurpassed and the geology is excellently exposed. Stop at designated viewing points to see the spectacular scenery and rocks in safety.
- **The Hoerikwaggo Trail:** This offers fine hiking up Chapman's Peak, through Silvermine Nature Reserve and along the trails of Table Mountain National Park; it brings you close to the rocks, and gives superb views.
- **Take part in spectacular road races:** Two of South Africa's biggest mass-participation events, the Cape Town Cycle Tour and the Two Oceans 56km-long Ultramarathon, follow Chapman's Peak Drive, so you can enjoy the geology as you participate in the sporting events.

3 Paarl Rock

PROVINCIAL HERITAGE SITE

Spectacular bare rock domes, formed by erosion of a small granite pluton

Paarl Rock seen from the vineyards of Laborie Estate in Paarl. Bretagne Rock is on the left, Paarl Rock is on the right and Gordon's Rock, the highest, is in the middle, largely obscured by the others. These large bald rock features are also called 'bornhardts'.

Parel, meaning 'pearl', was the name given by early Dutch settlers to the glistening domes that form a spectacular backdrop on the western side of the Berg River, and soon the same name was given to the historic town established in the area in about 1687.

Location and getting there

- The historic town of Paarl lies some 60km east-northeast of Cape Town, just off the N1. To get to Paarl Rock exit onto the R45 into Paarl and follow signs to the Paarl Mountain Nature Reserve. The outcrops are reached via the scenic Jan Philips Mountain Drive, accessed from the southern or northern sides of town.
- GPS (parking area): 33°44′34″S; 18°56′39″E

Setting the scene

Three landmark rock domes – classic examples of eroded granite and some of the largest such domes in the country – bulge out on the southeastern side of Paarl Mountain and overlook the scenic Berg River Valley and its famous vineyards. Directly above the town is the outcrop now called Paarl Rock (545m), and about 500m to the west are the more prominent domes named Gordon's Rock to the north and Bretagne (or Britannia) Rock (650m) to the south. In an effort to preserve its natural beauty and ecology, Paarl Mountain was proclaimed a National Monument in 1963 (now a Provincial Heritage Site).

Regional geology

Paarl Mountain is made up mostly of the oval-shaped **Paarl Granite Pluton**, which is more than 5km at its widest and 13km long. It is part of the **Cape Granite Suite** that intruded into tightly folded and metamorphosed sediments of the **Malmesbury Group** from 540–520 million years ago, at the start of the Cambrian. There are 10 other Cape granite batholiths and plutons in the southwestern Cape. The steep-sided shape of the intrusion is also called a 'boss' and it may be connected at depth to the Stellenbosch and Malmesbury batholiths.

Landscape and landforms

Paarl Mountain forms a broad topographic high that rises to 729m in the west at Groot Waboomkop. Preservation of the granite is due to its greater resistance to erosion compared to the more easily weathered surrounding meta-sedimentary rocks. The spectacular bare domes (also known as 'bornhardts') are superb examples of differential weathering and erosion shown by some massive granite bodies.

The granite bornhardts of Gordon's Rock (left) and Bretagne Rock (right background) seen from the western side at Nantes Dam. Note the obvious lack of major joints within the massive granite.

On Paarl Rock, remnants of a sheet of weathered granite, controlled by convex fractures parallel to the rock surface, reveal how the granite slowly breaks up. The now-isolated, rounded boulder is a residual corestone left after erosion. Bretagne Rock and Gordon's Rock are in the background.

Granite domes such as these result when large, curved, convex-upwards slabs and blocks break off along sheet fractures. These fractures are probably caused by relief of stress in the granite when erosion removes the considerable weight of overlying rock.

The large size of the domes is due to the wide spacing of vertical 'master joints' within the even-textured and rather structureless granite. There are fine examples of ribbed, gutter-like channels called 'rillen' and fracture-controlled clefts or 'kluftkarren', as well as small 'weathering pits' on the flatter rock surfaces.

Rock formations

The pluton is an excellent example of a multiple intrusion. In this instance, four phases of granite can be recognised, as illustrated in the geological cross-section (**Figure 10**). On the western side of the mountain is a quartz porphyry dyke that is not present in the area of Paarl Rock.

- Bretagne granite is the most extensive, forms the core of the pluton and was intruded first. It is greyish, coarse-grained, has prismatic feldspar phenocrysts, and occasionally shows small inclusions of recrystallised sedimentary rock.

On Bretagne Rock you can see unusual basin-like hollows in the granite, called 'weathering pits' or 'gnammas'. These hold rainwater and are formed by water-related weathering and erosion.

Thin dykes of fine-grained aplite cut through the granite at the end of magma intrusion and, because of the textural contrast with the coarse granite, they form negative, channel-like erosion features.

Typical texture of partly weathered Bretagne granite, consisting of prismatic feldspar phenocrysts set in a finer groundmass of quartz, feldspar and dark mica.

- Laborie granite occurs as an approximately 1km-wide zone on the eastern and northern sides, where it was intruded between the Bretagne granite and Malmesbury wallrocks. It is medium grained, consists of pale pink to white feldspar, colourless quartz and flaky dark mica (biotite). It is quarried at the De Hoop Quarry and known as 'Paarl Grey' in the trade.
- Montvue granite forms smaller bodies of darker, biotite-rich granite along the pluton margin in the east and northeast. It is interpreted as a 'hybrid' granite formed by the reaction of earlier granite magma and metamorphosed wallrock. The Montvue granite is limited in area and is hardly seen on the line of the cross-section; it is found only in small areas in the east and northwest.
- Bethel Dam granite occurs as small bodies in the centre of the pluton and to the northwest,

Figure 10: East-west geological cross-section across Paarl Mountain showing major rock formations and landmarks.

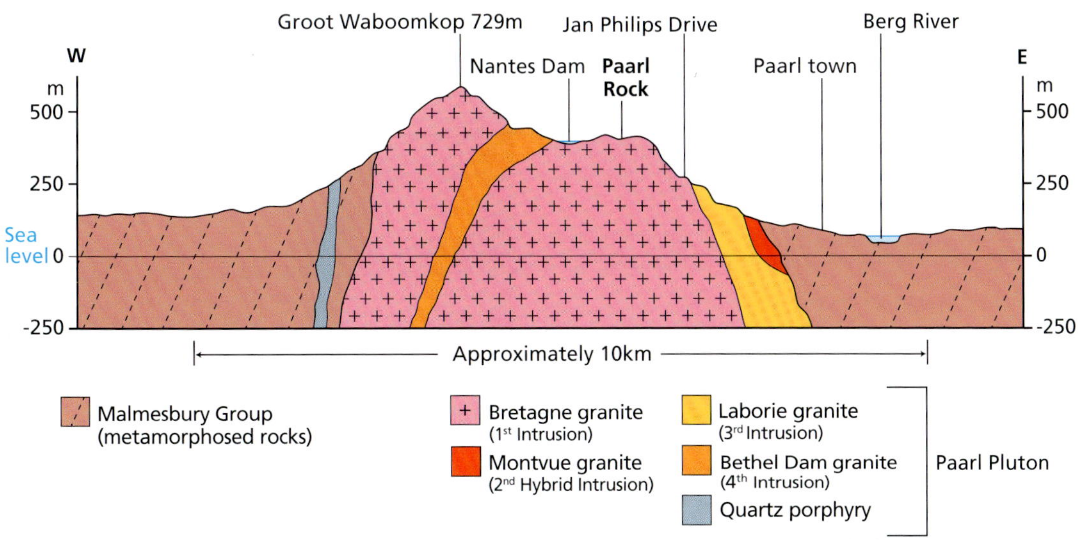

Adapted from information pamphlet Site C7 by the Western Cape Branch of the Geological Society of South Africa

and is richer in quartz and biotite. It was the last granite phase to intrude, as distinct, fine-grained, granite-filled conduits, except for later dykes of quartz porphyry and thin veins of aplite.

- On the western side of the pluton there is a low ridge of thermally metamorphosed Malmesbury shales (called 'hornfels') – but which is not part of the Paarl Rock geology.

At the top of Bretagne Rock, showing a weathering pit and a thin aplite dyke. Table Mountain sandstone forms Simonsberg in the far distance.

A complex geological story

Paarl Mountain has a long and complex geological story to tell, the events being broadly as follows:

- Granite magma was intruded in phases into strongly folded and metamorphosed sediments of the Malmesbury Group, about 8km below the Earth's surface and at a temperature of around 700°C. Heat of the intrusion created a wide zone of thermal metamorphism, forming metamorphosed rocks known as hornfels.
- The mountainland created by the Saldanian Orogeny was eroded down to around sea level over the next 50 million years, and subsequently formed the basement for the Cape Basin. The granite pluton may have been exposed. Deposition and lithification of the 8km-thick Cape Supergroup followed and the entire region was later involved in the Cape Orogeny, starting around 280 million years ago.
- Considerable crustal uplift ensued and over millions of years the overlying Cape Supergroup rocks, now seen only on surrounding mountains, were also removed by erosion. The Paarl Pluton was exposed and the rounded domes of granite that we see today were formed.

Things to see and do

- **Jan Philips Mountain Drive:** This 12km scenic route takes you along the eastern slope of the pluton; you need to turn off for the viewing sites and mountain trails.
- **The view from Paarl Rock:** You can drive to this viewing site where a descriptive plaque has been placed by the Geological Society of South Africa.
- **Visit the Nature Reserve:** Most of the mountain is covered by drivable mountain roads and hiking trails, including mountain bike tracks.
- **See the imposing Afrikaanse Taalmonument:** It's worth visiting this National Heritage Site, located at the southern end of the mountain. Paarl has a very strong cultural link to the Afrikaans language, which was proclaimed an official language in 1925.

The Afrikaanse Taalmonument on the southern slope of Paarl Mountain commemorates the founding of the Afrikaans language in 1925.

4 West Coast Fossil Park

NATIONAL HERITAGE SITE

One of the world's richest sites for Pliocene land fossils, here seen in place in their phosphate rock formation

Inside the area covering the archaeological dig, showing part of the prolific fossil bone bed that was once a phosphate mine. Using walkways, visitors can go on a conducted tour of the site, with posters also telling the dramatic story.

This world-renowned fossil site has an abundance of exceptionally well-preserved fossils reflecting the incredible assemblage of life that flourished here about 5 million years ago. The marketing line sums it up: 'When the West Coast was wild and the prehistoric Big Five roamed free'.

Location and getting there

- The Fossil Park is located near Langebaan Road, close to Saldanha Bay and the West Coast National Park. Take either the R27 north from Cape Town for about 120km and follow the signs; or from the N7 follow the R45 to Vredenburg a short distance north of Malmesbury.
- GPS (car park): 32°57′25″S; 18°07′04″E

Setting the scene

Fossil bones first came to light here in 1958, but it was in 1976, during the course of phosphate mining operations, that an amazing fossil bone bed was uncovered. Staff of the South African Museum rescued many fossils and discovered one of the richest accumulations of Pliocene fossils in the world. After the mine closed down in 1993 the Langebaanweg fossil site was declared a National Monument and is now a National Heritage Site. In 1998, assisted by the mining company, it was developed into the West Coast Fossil Park, managed by Iziko South African Museums.

Rehabilitation has been carried out at a quarry which is now a refuge for bird life and small mammals. Much of the vegetation is exotic rooikrans, which is progressively being removed.

Regional geology

The Fossil Park is located on the coastal plain along the Atlantic West Coast, a region that has repeatedly been subjected to changing sea levels during the Neogene Period (see p.24). It is thinly blanketed by young sedimentary rocks of the **Sandveld Group**, deposited from Miocene to Quaternary times, from 16 million to less than 1 million years ago. The fossil deposits are Pliocene in age, around 5.2 million years old. Millions of tons of phosphate-rich rock were mined to produce agricultural fertilizer from the then Varswater Quarry. This deposit is part of extensive sedimentary phosphate-bearing layers found in the Western Cape, formed earlier by oceanic upwelling of phosphate-rich sea water.

Landscape and landforms

The Fossil Park lies about 50m above sea level on a flat coastal terrace and is some 11km from the coast at Saldanha Bay. Locally, the highest point is the 70m-high Anyskop, about 1km south of the fossil beds. Earlier mining resulted in the formation of several extensive but shallow quarry areas and waste rock dumps.

Fossil Park Manager Pippa Haarhoff looking through gravel for micro-fossils at the outside sieves, where members of the public can also try their hand at fossil-seeking.

Rock formations

The relatively thin fossil-bearing sequence is called the Varswater Formation and comprises sediments showing repeated fluctuations in sea level, with depositional conditions varying from shallow marine to estuarine, marsh and fluvial. The detailed sedimentary history of these rocks is complex and includes southwest-trending river channels that host spectacular fossil-rich bone beds (or bone conglomerates) cut into the approximately 11m-thick phosphate rock layer. These deposits are overlain by calcified windblown sand (dune rock) of the much younger Langebaan Formation.

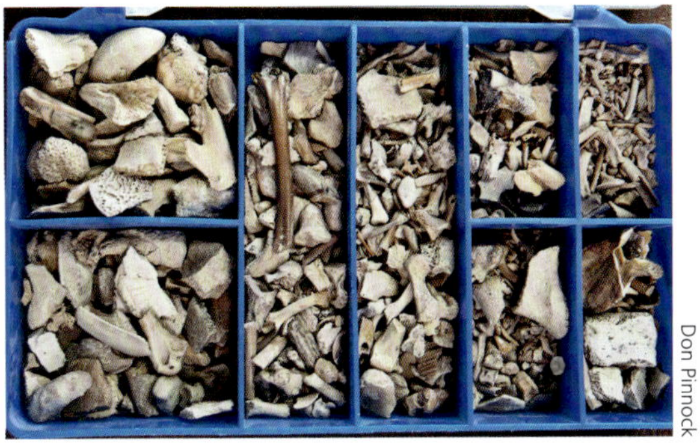

Don Pinnock

Countless fossil bones of small mammals, birds and amphibians have been found. This box contains, among other remains, tortoise shell, bones of francolin, hare, frog and snake, and teeth fragments.

The covered palaeontological dig sites and sorting area on the north side of the old Varswater Quarry. The famous bone bed is under the tunnel cover on the left.

The reconstructed fossil story

Around 5 million years ago, global warming and melting of the polar ice sheets resulted in a major rise in sea level, up to 90m higher than it is today, so that the coastline was much further inland. The Langebaanweg area is close to where the ancient Berg River flowed into Saldanha Bay. The coastal flood plain and a wide swampy estuary were bordered by subtropical riverine forest, with encroaching savanna, grassland and fynbos vegetation. The Berg River had cut a wide channel through existing soft, phosphate-rich rock close to a deeper eroded pool and sandbar. This was to become the place where thousands of animal bones collected and became fossilised.

During flash floods many of the larger animals living on the swampy coastal plain, notably hundreds of sivathere, were drowned by flood waters. Animal bodies were carried downstream to the estuary where they became beached or trapped against rock outcrops. Carcasses were scavenged by flesh-eaters such as hyaenas and bears, and bones accumulated in the deeper water and at the river's edge, where they were broken and trampled into the wet sand. During storms, carcasses of dead whales, seals and dolphins were carried into the estuary, where they became beached and where their bones accumulated.

Around 4 million years ago the inevitable cycle of climate change resulted in global cooling and lowering of sea level, and the environment became cooler and drier. The ancient Berg River changed its course northwards and many animal and plant species died out. The 'bone layer' became covered by about 20m of younger coastal sand and lay undisturbed until recent mining.

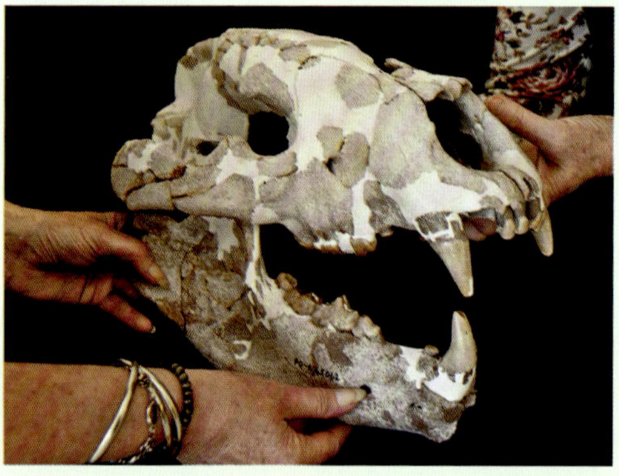

*Reconstructed skull and jaw of the African Bear (*Agriotherium africanum*), the only bear species ever found south of the Sahara.*

Some hippo teeth, although the species is not known.

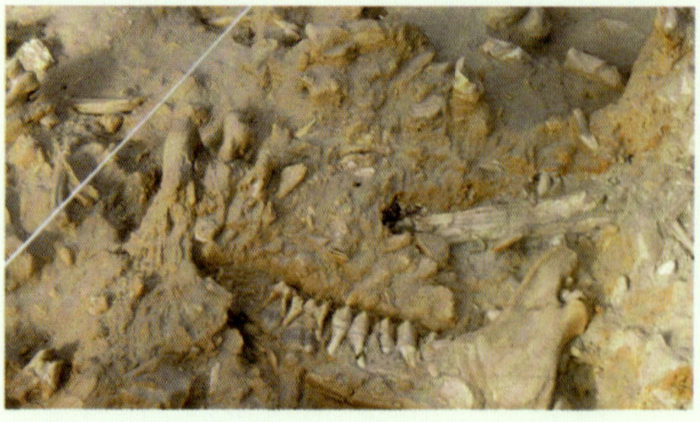

Juvenile sivathere jaw with teeth and ossicone fragment.

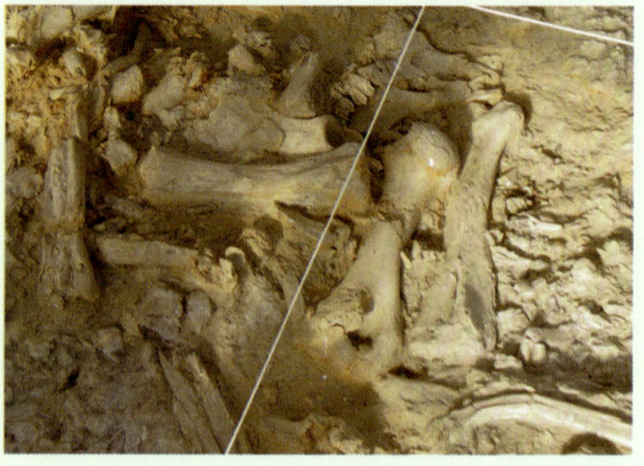

Juvenile sivathere tibia, plus pelvis, humerus and radius.

The countless fossils occur mainly within an exceptionally rich layer of jumbled bones; the lack of any complete animal skeletons suggests the area was subjected to catastrophic floods. While perhaps 80% of the fossils were lost at the mine site during its 40-year operation, considerable research has been done on the recovered fossils, mainly housed at Iziko South African Museum in Cape Town. More than 200 species of fossil, both vertebrate and invertebrate, have been identified, including over 80 species of large and small mammals, an exceptionally large collection of delicate bird bones, as well as reptiles, amphibians, seals and sharks.

*The western side of the quarry shows the contact between the older phosphate-rich, fossil-bearing Varswater Formation rocks (**1**) and overlying younger Langebaan Formation windblown sandstone (**2**) separated by a brownish zone of ancient soil formation (**3**).*

Things to see and do

- **The museum:** Located at the old mine offices, this provides information and some interesting displays, and in the laboratory you can see how fossils are prepared. A new Visitors Centre is planned.
- **The guided tour:** This is a must-do – a trained guide will take you to the covered dig site where you can see countless partly excavated fossil bones in place in their host rock. Look out for the graphic posters that tell the dramatic story of the fossils.
- **Your own dig:** Outside the dig site people can try their hand at recovering small mammal, bird and amphibian fossils from the soft phosphate sand, to add to the collection.
- **Other activities:** You can go on mountain bike trails and wildflower rambles, go horse-riding or bird-watching.
- **Langebaan Lagoon:** Some 15km to the west of the Fossil Park is the West Coast National Park, with its own geological attractions, ranging from excellent exposures of Cape granite to very much younger windblown coastal sandstones. One such attraction is at Kraal Bay on the western shore of the lagoon, where the well-preserved 'Eve's footprints' were found in coastal sandstone, believed to be around 120,000 years old. There is a new Geological Society plaque at Seeberg (Site C8).

5 Stadsaal Caves

CEDERBERG CONSERVANCY

Eroded sandstone formations in a mountain wilderness,
plus an amazing display of San rock art

The pathway leading to the main Stadsaal Caves entrance: the rocks are part of the Nardouw Subgroup sandstone sequence, the upper part of the Table Mountain Group. Bizarre rock formations – and the sandy floor – have been created mainly by wind erosion.

The spectacular Cederberg abounds with strange-looking, brightly coloured eroded rock formations, making a landscape like no other in South Africa. The Stadsaal Caves, with their craggy rock faces, weathered columns, isolated pinnacles and gargoyle-like formations, make for an intriguing geological site.

Location and getting there

- Stadsaal Caves are located off the beaten track within the Matjiesrivier Nature Reserve, close to the gravel road through the Driehoeks River Valley. The best routes into these rugged mountains are either off the N7 north of Citrusdal and via the forestry station at Algeria; or northwards from Ceres on the R303, then via the Grootrivier Pass to Matjiesrivier. Permits are required to visit specific rock sites and hiking trails.

- GPS (Stadsaal entrance): 32°31′19″S; 19°19′01″E

Setting the scene

The Cederberg has long been a favourite wilderness area, well known for its many hikes, spectacular scenery and mountain fynbos and succulent Karoo vegetation. The name comes from the endemic and threatened Clanwilliam cedar tree that was once abundant on the high mountain slopes. The Cederberg Conservancy covers most of the region and includes the Cederberg Wilderness Area, proclaimed in 1973 – the first such wilderness in South Africa. This is the northernmost area of the Cape Floral Region's World Heritage Site, and is managed by CapeNature.

Regional geology

The Cederberg is part of the western branch of the Cape Fold Belt where roughly north-aligned deformation of the Cape Fold Belt was less intense, and now forms a broad, anticlinal structure. The rock formations at Stadsaal Caves are in the upper part of the **Table Mountain Group**, lying above the Cederberg shale marker unit. These sandstone strata dip gently eastwards and are overlain by the **Bokkeveld Group** in the east.

Another view of part of Stadsaal Caves, in this case a series of peculiar sandstone columns that have formed by differential weathering and wind erosion, and are related to strong vertical jointing. Notice the remains on the roof of columns that have eroded away.

Landscape and landforms

The Cederberg is considered to be a remnant of the ancient surface above the African land surface (see pp.36–37), thus indicating that erosion has taken place over many millions of years. Erosion starts with weathering along countless vertical joint surfaces within gently dipping sandstone layers, coupled with the eroding action of wind and water. It's worth looking at Google Earth, where images reveal the intensity and extent of rock jointing. The bizarrely shaped and brightly coloured rock formations are a perfect demonstration of natural sculpting, and testimony to the resistance of rock to the power of erosion.

San rock art

The hunter-gatherer San people migrated through the Cederberg for many thousands of years before European colonisation took over in the 18th century, and they left an amazing legacy of rock art. It is said that the Cederberg region has more rock paintings per square kilometre than anywhere else in southern Africa. Despite the ravages of time and weather, many of the paintings have lasted for thousands of years. Images are varied: some show humans with various animal features; others show hunting scenes or processions of people, or scenes of animals. It is now generally accepted that San rock art was not decorative, and that the paintings should not be interpreted literally: they conveyed important spiritual meaning for the San people, particularly for their shamans.

A cave-like overhang in eroded sandstone has protected the famous 'elephant rock' painting for at least 1,000 years; strangely, elephants did not live in this area. This heritage site is located about 1km north of Stadsaal Caves. Inset: A close-up of the elephant group.

The reddish to orange colours are due to oxidation of iron minerals and to abundant lichen on the rock surfaces.

This is the result of ongoing erosion by wind of a less resistant sandstone pillar near ground level. Eventually the pillar will be worn away completely near the base and the upper part will collapse.

Rock formations

The Stadsaal Caves, with their weird and wonderful natural rock sculptures, have been wind-weathered and eroded out of well-jointed, brownish sandstones of the Nardouw Subgroup, the upper part of the thick Table Mountain Group. These sandstones were laid down around 440 million years ago as quartz-rich sands on a braided river plain or on a shallow marine shelf. Sedimentary layering, both planar and cross-bedded, as well as gritty and pebbly layers, are common indicators of shallow-water deposition. This took place long before the break-up of Gondwana within a continent-wide, east-west-trending Cape Basin in which some 8,000m of sediment would eventually accumulate. Later, the entire sequence was mildly folded during the Cape Orogeny, starting around 280 million years ago.

A close-up view of a quartz pebble layer in the gritty sandstone, indicating that the quartz-rich sediment was deposited in shallow water, probably an alluvial braided river plain. The reddish coloration is largely from lichen.

Things to see and do

- **Explore Stadsaal Caves:** These caves provide a wonderful experience as you make your way around and into the numerous eroded formations. Look out for the historic graffiti, dating from the 19th century, in the main cave.
- **See the San rock art:** The well-known 'elephant scene' is found about 1km north of Stadsaal in another similar mass of eroded sandstone. Enquire locally about other sites open to the public (permits are required). One of the best private areas is Bushman's Kloof Resort north of Wuppertal.
- **Visit Truitjieskraal:** Located only 2.5km to the south of Stadsaal but accessed on a different road, this is another labyrinthine pile of wind-eroded sandstone, also noted for its San rock art.
- **See the Maltese Cross and Lot's Wife:** The Maltese Cross is a massive pillar of jointed sandstone, an isolated remnant of a once continuous sheet. Located west of Stadsaal, it requires a hike to view it. Another extraordinary rock formation, known as Lot's Wife, is near Dwarsrivier and a short walk off the gravel road.
- **Climb to Wolfberg Cracks and Wolfberg Arch:** To the northwest of Stadsaal, on high ground and after a not-so-easy climb via the spectacular Wolfberg Cracks, are a large natural arch and other remnants of wind-eroded sandstone. There are many other hikes in the region.

6 Cogmans Kloof

A LEGACY OF THOMAS BAIN, ROAD-BUILDER

Magnificently folded sandstone strata in a deep poort that cuts through the Langeberg

The eastern side of Cogmans Kloof from the old English Fort, as one looks northeast towards Montagu. The tarred road is on a low-level river terrace adjacent to the Kingna River, which periodically floods. The mountain slope in the distance shows an array of intense tight folding, the precise origin of which is still debated.

The name 'Cogmans' can be traced back to a local Khoikhoi clan called the Kockemans or Kogmans that lived in the area in the 18th century. Also known as Kogmanskloof, this spectacular route through the Langeberg range follows the course of the Kingna River, a tributary of the Breede River.

Location and getting there

- The 4km-long poort is located at the start of Route 62, the easiest and best road into the Little Karoo, and it is a busy thoroughfare. It lies between the towns of Ashton to the southwest and picturesque Montagu to the northeast, also known for its hot spring.
- GPS (Tunnel parking area): 33°48′11″S; 20°05′44″E

Setting the scene

In the early 18th century a rough track following the Kingna River bed was the only way through the almost impenetrable Langeberg, and it was subjected to regular flooding. Master road-builder Thomas Bain was called in; he got to work in 1875 and the pass was finally opened in 1877. A 16m-long tunnel had to be blasted through the massive sandstone barrier known as Kalkoenkrans, and dynamite was reputedly used here for this purpose for the first time in South Africa. Bain's narrow old road still exists in places along the eastern side of the poort.

Regional geology

The impressively folded rocks seen in this spectacular poort are almost entirely of the **Table Mountain Group** – the lower part of the **Cape Supergroup** – and they were strongly folded during the Cape Orogeny. On the Ashton side there are exposures of the older **Malmesbury Group** metamorphosed shales and a major crustal dislocation known as the Worcester Fault.

Landscape and landforms

The Langeberg is one of the Cape's major mountain ranges and is a remnant of ancient mountainous terrain that is found above the African land surface (pp.36–37). Although named Cogmans *Kloof*, this spectacular product of river erosion is more correctly known as a 'poort'. Given that at a much earlier time the Cape mountains were of Himalayan proportions and probably reached several thousand metres in height, it is astounding that the relatively minor Kingna River could have caused such substantial erosion. Geomorphologists are not agreed on how the river cut down through the very hard, folded sandstone layers and over what period of time this happened. Explanations include the down-cutting of an earlier drainage (possible incised meanders

Folding and faulting

Here you can see folding of the Cape Fold Mountains at its very best, particularly in the intensely folded zone in the mountain mass as you head towards Montagu. The folding was due to a period of very strong compression within the crust during the Cape Orogeny (see Western Cape Overview, pp.38–39). The overall geological structure is of a large anticline in mainly Peninsula sandstone on the Ashton side, with smaller subsidiary folds, and a major syncline in Nardouw sandstone on the Montagu side (see cross-section). The syncline includes the intensely folded zone that has been attributed to collapse under the influence of gravity. A major zone of crustal dislocation, the important northwest-trending Worcester Fault, occurs close to Ashton. To the south of the fault all formations have dropped down by several kilometres – a result of the break-up of Gondwana.

Figure 11: A simplified southwest-northeast geological cross-section through Cogmans Kloof.

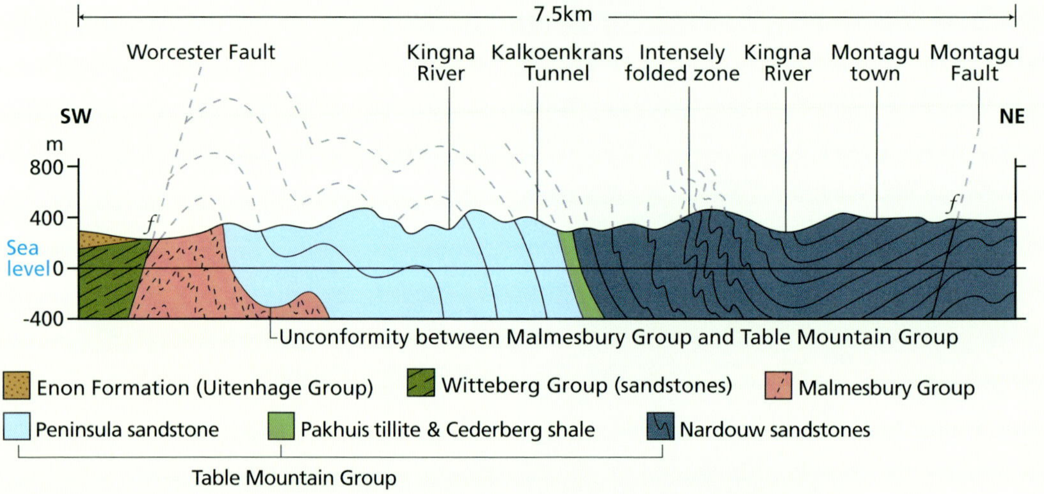

Adapted from information pamphlet Site C10 by the Western Cape Branch of the Geological Society of South Africa

Thomas Bain's original tunnel through the very hard sandstone at Kalkoenkrans. On the koppie above the tunnel is a small blockhouse, also known as Fort Sidney, built of local stone in 1899 during the Anglo-Boer War.

exist) or possibly strong headward erosion from ravines on both sides of the mountain, followed by the capture of one by the other. A much more recent landform is the narrow river terrace, several metres higher than today's river, which provides the base for the roads, old and new.

Rock formations

- The dominant formations are hard Table Mountain Group sandstones: light grey, thickly bedded Peninsula Formation sandstone (as on Table Mountain, see pp.42–48) on the southwestern (Ashton) side, followed upwards by brownish-weathering, thinner-bedded Nardouw Subgroup sandstones (as in the Cederberg, see pp.64–67) on the northeastern (Montagu) side. But in contrast to Table Mountain where the beds are essentially horizontal, here they are folded and, as a result, are steeply dipping.
- Northeast of the tunnel, forming a steep valley between the sandstone sequences, is a narrow poorly outcropping zone of Pakhuis Formation tillite and Cederberg Formation shale which forms an excellent regional marker.
- At the southern end of the poort, close to Ashton, steeply dipping Malmesbury Group phyllites (metamorphosed shales) are exposed along the roadside and show the folded basement formation upon which the Peninsula sandstone was deposited.

Thomas Bain's original old road on the eastern side of the river near the tunnel is still in reasonable condition nearly 140 years later.

Near the tunnel you can see outcrops of layers of hard Peninsula Formation sandstone, part of the Table Mountain Group. Now almost vertical in attitude, these were once horizontal beds, intensely folded and upturned during the Cape Orogeny.

Things to see and do

- **Drive through the poort:** A slow drive between Ashton and Montagu, with a stop at the tunnel, is a geologically exhilarating experience. A Geological Society plaque at the tunnel parking area describes the local geology.
- **See Thomas Bain's old road:** With its signature stone walling, this can be seen in two places: at the bridge on the Ashton side of the tunnel, and close to another road bridge about 1.4km further back towards Ashton, where the old road hugs the cliff along the east bank of the river. There is a 1974 National Monuments Council plaque in Bain's honour at the tunnel parking area.
- **Climb to the English Fort:** Steps at the tunnel lead up to the blockhouse on top of Kalkoenkrans. It was built by the British in 1899 at the start of the Anglo-Boer War, using local stone. It was part of the local defences to stop any incursion by Boer forces, and it now provides splendid views.

SOUTHERN AND EASTERN CAPE

OVERVIEW OF LANDSCAPES AND GEOLOGY

This chapter focuses on nine geological sites in the southern part of the Western Cape (Geosites 7–11) and in the Eastern Cape (Geosites 12–15). As shown on the national maps (pp.12–13), these are mostly associated with the Cape Fold Belt or with the southern Karoo, and their distribution on the regional landscape and geological maps (pp.74–75) reveals the very linear nature of the fold belt (although Geosites 14 and 15 are off these particular maps). Relevant geological formations range from the Late Proterozoic to much younger Cretaceous deposits.

The major geological formations of the region are listed below:

■ Cango Caves Group
(Late Proterozoic)

This forms an elongated inlier along the southern foothills of the Swartberg range, roughly between Calitzdorp and De Rust. They are the oldest sedimentary rocks in the region. Deposited around 600 million years ago, they are similar in origin and deformation to the **Malmesbury Group** of the Western Cape, created during the formation of the Gondwana supercontinent. They comprise a layered sedimentary sequence of metamorphosed clastic and carbonate rocks, all dipping steeply, and providing the basement upon which the sandstones of the **Cape Supergroup** were deposited. In the area north of Oudtshoorn these rocks often stand out as elongated, bush-covered limestone hills showing karstic features, the most notable being the famous Cango Caves (**Geosite 8**) and also the strongly sheared formation that is seen at the southern entrance to Meiringspoort (**Geosite 9**).

■ Cape Supergroup
(Ordovician to Carboniferous)

The formations that make up this supergroup continue eastwards from the Western Cape, essentially unchanged, as the same **Table Mountain**, **Bokkeveld** and **Witteberg Groups** (see Overview of the Western Cape). In the southern Cape, particularly in the poorts and passes, the full geological sequence of the entire supergroup is excellently displayed, and again it is the thick and resistant Table Mountain quartzitic sandstone that dominates the landscape, and makes the spectacular coastal (Outeniqua and Tsitsikamma) and inland (Great Swartberg and Grootwinterhoek) ranges. The dramatic folding of the strata, seen in the Swartberg Pass (**Geosite 7**) and Meiringspoort (**Geosite 9**) stands out most vividly. These are the remains of an ancient Himalayan-proportioned mountain range (see Cape mountain building, p.39) and in places still reach over 2,000m, with remnants of an ancient Gondwana land surface. The Cape mountains continue eastwards, eventually decreasing in elevation and grandeur. Surprisingly, these are the same formations that make up much of the precipitous coastline of the southern Cape, which was planed down by the sea during the Tertiary, and subsequently uplifted to produce a flat coastal platform. The Knysna Heads (**Geosite 10**) and Storms River Gorge (**Geosite 12**) are comprised of hard Table Mountain quartzitic sandstone.

Southern foothills of the Swartberg on the road to Swartberg Pass. The low rounded hills in the middle distance are folded sediments of the older Cango Caves Group; the high mountains are of folded Table Mountain Group quartzitic sandstone.

Karoo Supergroup
(Carboniferous to Jurassic)

North of the Cape Fold Mountains, the large Karoo Basin formed as a result of sagging of the crust in direct response to the mountain building to the south. The basin continued to subside steadily under the weight of sediment from the rapidly eroding Cape mountains being deposited within it, along with fine ash that came from volcanic activity far to the south, which produced thin layers of tuff. The glacial **Dwyka Group**, the deeper water **Ecca Group**, and the lower part of the **Beaufort Group** are found here, and were also caught up by Cape folding. The thick Beaufort Group, world renowned for its abundant prehistoric reptile life (see **Geosite 50**, Fraserburg Fossil Surface), was mostly formed by mud that was deposited on vast alluvial flood plains. Today the iconic Karoo landscape of flat-topped koppies, endless plains and rugged escarpments is an expression of the underlying geology and ongoing processes of denudation, in which intrusions of Karoo dolerite play an important role.

Karoo Dolerites
(Jurassic)

Around 183 million years ago, before the break-up of Gondwana, an ernormous amount of basaltic magma found its way to the surface where it erupted as the thick pile of **Drakensberg Group** lavas, and it also intruded the flat-lying Karoo sediments to produce innumerable dolerite dykes and sills. Because of their overwhelming abundance, these countless dolerite intrusions are not shown on the national geological map (**Figure 2**). The impressive southern Great Escarpment is largely a product of headward erosion meeting resistant dolerite sills. The Valley of Desolation (**Geosite 14**) and the Hole in the Wall (**Geosite 15**) are examples of erosion-related dolerite landforms.

Uitenhage Group
(Jurassic to Cretaceous)

The break-up of Gondwana resulted in a new faulted South African coastline developing. To the south, from around 120 million years ago, sideways movement along a major crustal fracture led to a continental fragment (the Falklands Plateau) drifting off with South America. This movement produced wrenching in the existing linear fold mountains, and resulted in east-west-trending rifts, forming a series of fault-bounded, half-graben valleys into which the sea partly flooded. Sediment derived from the rapidly eroding Cape Fold Mountains, and initially particularly thick and often reddish **Enon Formation** conglomerates, accumulated in these fault-controlled rift valleys, which today make several small elongated sedimentary basins. These are well displayed in the Little Karoo and in the conglomerates of the Gamtoos Valley (**Geosite 13**); and similar quartzite pebble conglomerates are seen at Robberg Peninsula (**Geosite 11**).

Figure 12: Landscape map of parts of the southern Cape and Eastern Cape. The linear mountains, in reddish-brown tones, dominate, a reflection of the east-west-trending Cape Fold Belt. The terrain is represented by high mountain ranges and wide valleys, including the Little Karoo and part of the more elevated brownish Great Karoo to the north.

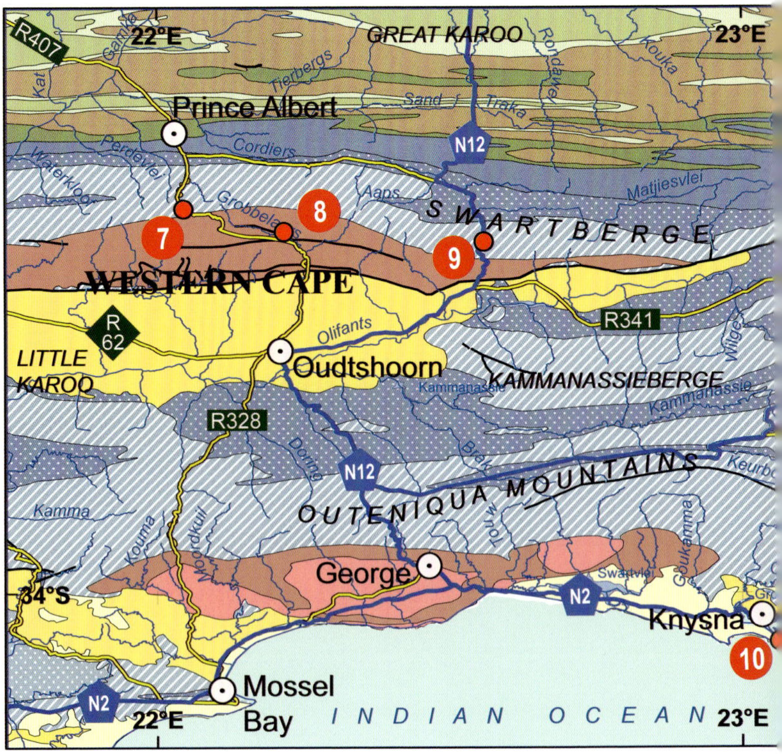

Figure 13: Geological map of parts of the southern Cape and Eastern Cape. The Cape Supergroup makes up most of the geology, in particular quartzites of the Table Mountain Group, and Karoo sediments in the north. Cretaceous sedimentary rocks fill younger half-graben rift valleys. The oldest rocks are the metamorphosed Cango Caves Group that lie to the south of the Swartberg, and which are related to the elongated Kaaimans Group, and Cape granite intrusions in the coastal George area.

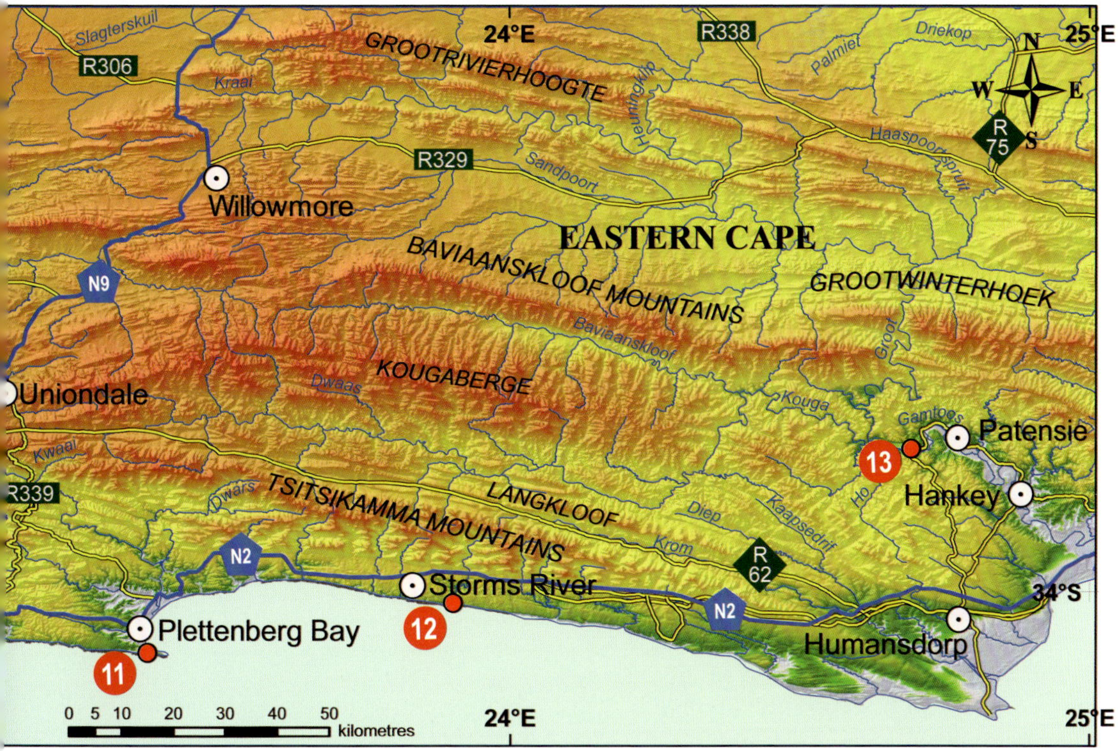

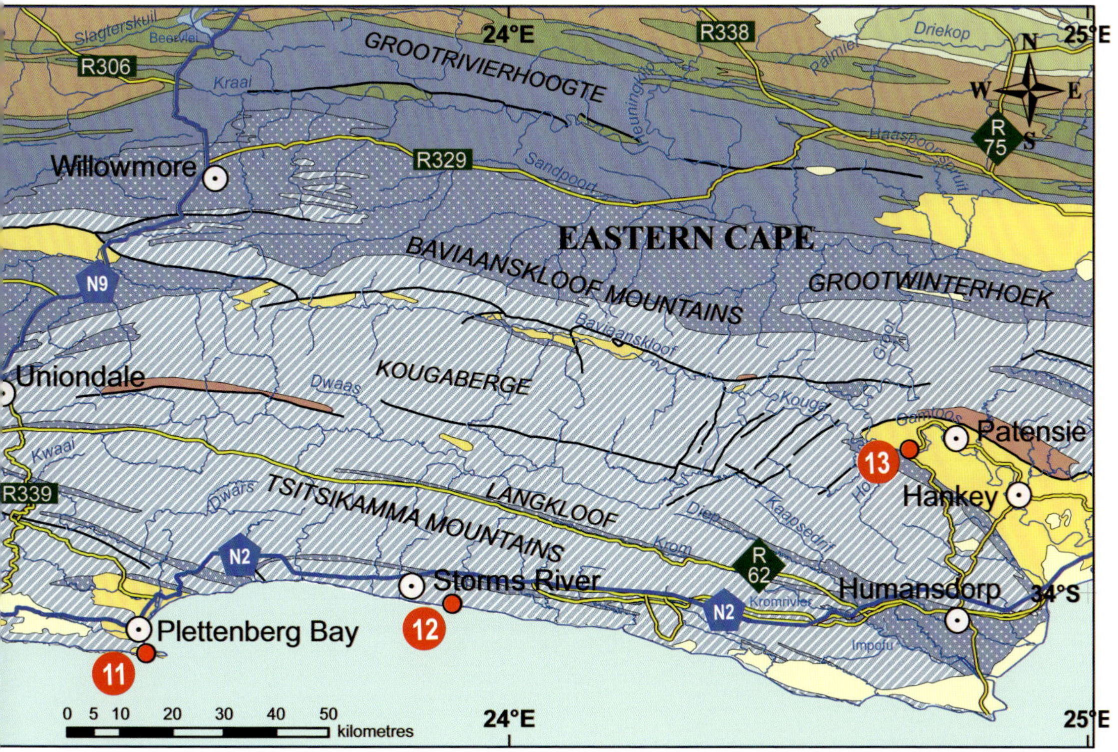

• A stratigraphic key to this simplified regional geological map appears on the inside covers

7

Swartberg Pass

PROVINCIAL HERITAGE SITE

A world-class mountain pass, an extraordinary demonstration of folded strata and an impressively eroding fold mountain chain

Near Tweedewater, close to the north end of the pass, intensely folded Nardouw Subgroup quartzitic sandstone of the Table Mountain Group dips vertically, proof of the immense Earth forces that operated at the time of the Cape mountain building. Nearby, look out for the isolated cross-like structure formed by jointing in the sandstone (see p.79).

The Great Swartberg forms an almost impenetrable mountain barrier between the Great Karoo and the Little Karoo, which are connected in only a few places. One of these is the awe-inspiring 24km-long Swartberg Pass, regarded as one of the most spectacular mountain roads in the world.

■ Location and getting there

- The Swartberg Pass links picturesque Prince Albert in the north and historic Oudtshoorn in the south, and straddles the Great Swartberg range. The road is untarred, reaches an altitude of 1,585m and is potentially hazardous in bad weather. The mountains make up the pristine, diversity-rich Swartberg Nature Reserve, part of the Cape Floral Region World Heritage Site.
- GPS (at The Top): 33°21′09″S; 22°02′47″E

Setting the scene

In the 19th century a reliable route into the Great Karoo was becoming essential as Meiringspoort was often impassable. Master road-builder Thomas Bain surveyed the route for the pass in 1875 and road building from the northern side started in 1881. He personally took over road construction in 1883, using about 240 convicts as manual labour and, after toiling through untold hardship in heat, rain and snow, completed the pass. It was formally opened in 1888. One hundred years later this magnificent endeavour was declared a National Monument, now a Provincial Heritage Site.

Regional geology

The entire sedimentary sequence of the **Cape Supergroup** is seen as you traverse this route. The **Table Mountain Group** makes up most of the pass, while lesser ridges and valleys formed by the **Bokkeveld** and **Witteberg Groups** appear at the Prince Albert end.

Landscape and landforms

The Great Swartberg is over 200km long, up to 25km wide and, in places, exceeds 2,000m in height. The extremely rugged terrain is dissected by countless ravines, and it is being eroded by the extreme climate. The steep south-dipping shape of the southern mountain face is a result of steeply dipping, overturned quartzitic sandstone beds. From the top, a surprisingly level upper surface is seen along the crest of the Great Swartberg range reminding one that the Cape Fold Mountains are remnants of an ancient land surface that lies above the old African erosion surface. On the northern side, the narrow gorge of the Dorps River has cut deeply into the quartzitic sandstones.

The hidden valley of Gamkaskloof, aka *Die Hel*

In about 1830 Boer settlers moved into a narrow, secluded, 20km-long side valley of the Gamka River deep in the Swartberg and lived in semi-isolation for around 130 years, undisturbed by progress and well sustained by subsistence farming. These resilient people have a fascinating history of hardship, endurance and self-sufficiency. Geologically the valley is formed by an elongated synclinal structure underlain by softer Bokkeveld shales and hemmed in by mountains of Table Mountain quartzitic sandstone. Access into the valley was by bridle path only, but in 1962 a 52km-long mountain road was finally constructed from the top of the Swartberg Pass. This signalled the end for the close-knit community as people slowly moved out, lured by the outside world – and as inquisitve outsiders came to see what had become an almost legendary valley.

Towards the northern end of the pass, in the gorge made by the Dorps River near Malvadraai, extraordinary complex folding of the Nardouw Subgroup quartzitic sandstones, the upper part of the Table Mountain Group, is very clearly seen.

Rock formations

- Thick, resistant Table Mountain Group quartzitic sandstones form the high ground. The thin Cederberg shale unit (see cross-section) separates the Peninsula Formation and Nardouw Subgroup that make up the majority of the range (see **Figure 14**).
- On the northern, Prince Albert side the road winds its way up through rugged terrain underlain by extraordinarily folded Nardouw quartzitic sandstones.
- On the precipitous south-facing mountain slopes the road snakes its way down Peninsula quartzitic sandstones, the same formation seen – albeit unfolded – on Table Mountain (pp.42–48).
- In the foothills at the southern end of the pass, there are poorly exposed and older rocks of the Cango Caves Group.

Mountain building

Excellent geological evidence can be seen here of the east-west-trending Cape Fold Belt

A close-up view of the steeply dipping beds of overturned Table Mountain quartzitic sandstone seen at the Skelmdraai viewsite on the climb up the southern part of the pass.

Around Tweedewater the folded quartzitic sandstone layers dip vertically, proof of the immense movement of the strata. Look out for the cross-like structure formed by jointing.

that formed during the Cape Orogeny around 280–230 million years ago (see Western Cape Overview, pp.38–39). The pass traverses the northern side of what has been geologically interpreted as a mega-antiform, a 40km-wide and 10km-thick complex fold structure that is fault-bounded in the Little Karoo to the south, and which dies out northwards in the Great Karoo. Part of this is shown in the simplified diagram below, illustrating part of a large complex fold structure, complicated by smaller folds, faults and thrusts. What is seen today is only a remnant of the folded strata that produced a mountain range of Himalayan proportions, now worn down by millions of years of erosion.

Figure 14: An idealised north-south geological cross-section through the Swartberg Pass, which is part of a large complex fold structure (much simplified, after IW Hälbich, 1983).

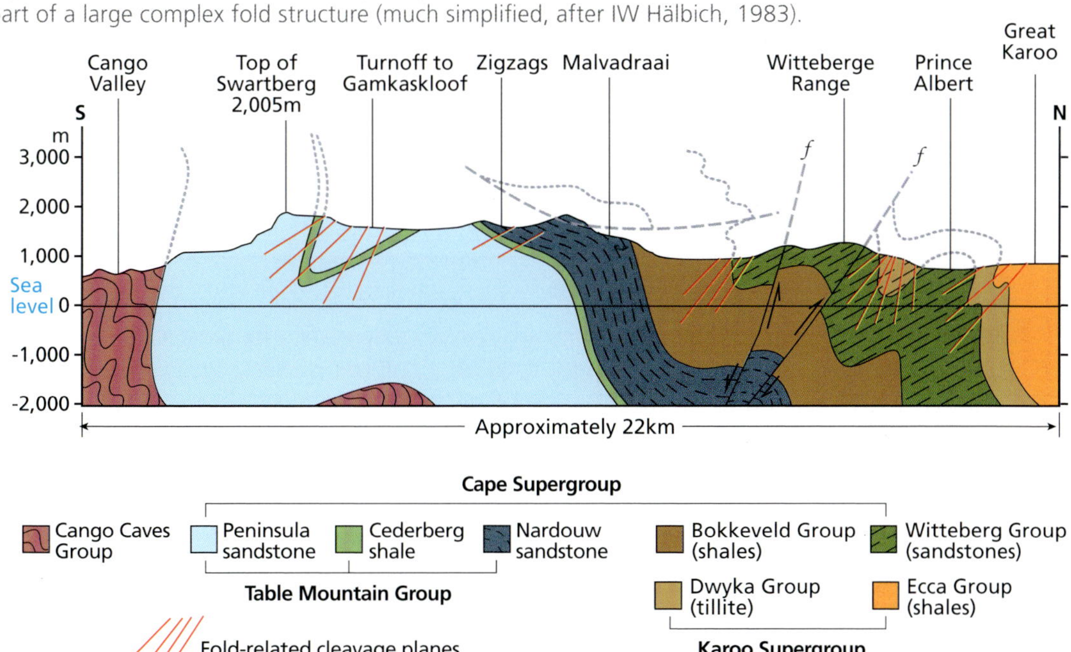

Adapted from IW Hälbich in *Geodynamics of the Cape Fold Belt*, Geological Society of South Africa (1983)

The famous zigzag part of the pass, seen looking northwards. The old jail-house (blikstasie) *was at bottom right.*

At the Prince Albert-side entrance to the pass, Witteberg quartzitic sandstone makes its first appearance, dipping moderately to the south and overlying typical Dwyka tillite.

The highest point of the south-dipping Great Swartberg in this area is an unnamed peak reaching 2,005m, directly to the west of The Top viewsite at around 1,585m. Notice the overturned, south-dipping rock strata.

This precipitous, 50km-long gravel road is not for the faint-hearted but takes you to the hidden valley, Die Hel, which was once inhabited by hardy Boer families.

Things to see and do

- **Drive the pass:** It is impressive travelling in either direction, from Oudtshoorn or from Prince Albert. Stop at The Top viewing site to appreciate vistas across the Matjiesrivier Valley and far beyond, and look out for the different landforms. Looking northwards, you can see the distant Great Karoo.
- **Experience Gamkaskloof:** There is a turn-off to Gamkaskloof in the middle of the pass but the 52km-long road into this hidden valley, also known as Die Hel, is not for the faint-hearted; a 4x4 is not essential, but high ground clearance is good. The kloof has become a tourism highlight, and accommodation needs to be arranged with CapeNature.
- **Hiking and 4x4 trails:** Those who are more active can enjoy several spectacular trails starting at the Ou Tol on top of the Swartberg, all with unsurpassed views and overnight facilities.
- **Visit Prince Albert:** This delightful, stylish old village dates back to 1762 when European settlers started farming in the sunny and productive valley. Visit the Fransie Pienaar Museum to see the Roy Oosthuizen fossil collection and to learn about the minor gold rush that took place at nearby Prince Albert goldfield in the 1890s.

8

Cango Caves

PROVINCIAL HERITAGE SITE

A spectacular limestone cave system that is South Africa's finest show cave and one of the country's best-known natural wonders

South African Tourism

A scene from Botha's Hall, the second of several massive caverns in Cango I. Shown here are drape-like limestone features known as The Curtains and a joined-up stalagmite and stalactite, forming a 13m-high 'completed' column.

In 1790, after the discovery of a cave opening, local farmer Jacobus van Zyl was the first person to be lowered into what is now called Van Zyl's Hall. By the light of flickering flames the early pioneers were amazed to find a vast cavern filled with pristine cave formations. This is South Africa's premier, most spectacular cave.

Location and getting there

- The Cango Caves are found in the undulating foothills of the Great Swartberg some 30km north of Oudtshoorn. You get there on the R328, the same road that takes you up the Swartberg Pass. The road is tarred and there is considerable tourist infrastructure on site.
- GPS (entrance to main building): 33°23′34″S; 22°12′53″E

The hillside entrance to the Cango Caves – the opening discovered in 1790 by Jacobus van Zyl.

Setting the scene

By the early 19th century the caves were becoming a growing attraction, which led to their partial interior destruction. This prompted regulations governing their operation to be

Fine-grained, medium grey limestone of the Matjies River Formation, dipping approximately 70° south, forms the host rock to the Cango Caves, which are a much more recent karst development probably dating from the late Tertiary.

The limestone hills around the Cango Caves have other caves as well. Inset: Nearby, this bush-covered hill west of the Caves reveals a cavernous karstic opening.

issued in 1820, including an entrance fee and a ban on removing cave formations. Archaeological work has shown that human activities in the caves go back to the Stone Age and that the entrance was used by the San and decorated with their rock paintings. The caves were declared a National Monument in 1938 and now a Provincial Heritage Site.

Regional geology

The caves are found in the **Cango Caves Group**, an east-west-trending inlier of Late Proterozoic age that is best seen in nearby Schoemanspoort. This elongated inlier, some 140km long and up to 14km wide, is exposed on the southern side of the Swartberg and is bounded to the south by the major Cango Fault.

Landscape and landforms

The limestone foothills of the Great Swartberg reveal little on the surface to indicate the existence of caves, which were formed by an underground stream diverted from the surface, and slow chemical erosion of limestone: a karst-forming process (see text box, p.84) related to shallow ground-water movement dating from the Tertiary or even later. Today, a perennial stream flows about 15m below the present cave, in a much smaller passage.

The Cango system of tunnels and its many named chambers is known to extend underground for at least 5km and comprises Cango I, Cango II and Cango III–IV. The system may be even longer, and other, as yet undiscovered entrances may exist somewhere in the limestone hills. The upper caverns are essentially dry and contain spectacular and varied displays of limestone cave formations, and many side chambers. Cango I is the only

How limestone and dolomite caves are formed

Caves such as these, found in limestone rocks, are called 'karst structures', the name coming from the Karst Plateau close to the Adriatic Sea, which has an abundance of limestone caves. The essential process of cave formation, as shown in the diagram, is the result of very slightly acidic rainwater penetrating along vertical fractures, joints and bedding planes, very slowly dissolving the surrounding carbonate rock (limestone or dolomite) and forming water-filled caverns (phreatic cavities). This takes place below a stable land surface where dissolution is assisted by good rainfall. In places, the cavities join up and provide a pathway for running underground water. As the level of the water table drops (or uplift takes place) the cavities are drained and become exposed to air (vadose cavities), and much younger secondary cave formations, generally known as 'speleothems', are formed from mineral-rich dripping water.

Figure 15: How a solution cave system is formed in limestone rocks over a long period of time.

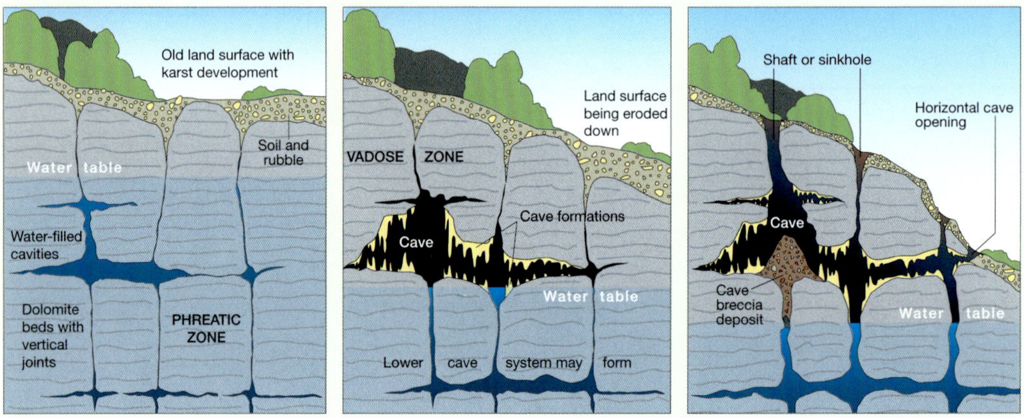

cave open to the public and is 775m long and fairly level. In 1972 a small passage was widened, leading to the pristine Cango II, now called the Wonder Cave, which extends a further 270m and is filled with spectacular cave formations. Cango III was discovered in 1975 by following a deeper underwater passage, which revealed another 1,380m-long cave.

Rock and cave formations

The caves are hosted in pure grey marine limestone, which is part of the Matjies River Formation. This mixed shale and limestone formation is 20km long, 2km wide and dips steeply to the south. Clean fresh rocks are not seen in the cave because of younger cave formations, as well as poor light, but are exposed on the hill outside. Of major interest are the spectacular interior cave formations, many of which have been given popular names. They formed over countless years by dripping water, and are of the travertine variety of pure calcite, which, when broken, shows extremely fine banding. Their coloration is due to slight contamination by iron or manganese. In the first part of the caves, Cango I, the cave formations are no longer forming: they are mostly 'dead' and partly corroded. But in the deeper parts (Cango II onwards), where there has been little human access, formations continue to grow, and spectacular pristine cave formations still exist.

Things to see and do

- **Do the Heritage or Adventure guided tours:** The first lasts an easy 60 minutes and takes you through the well-illuminated front part of Cango I, where you will see spectacular cave formations. The latter takes 90 minutes, is more demanding and gets you into and through some restricted spaces.
- **Visit the Interpretation Centre:** This will give you a greater understanding of the geology and history of the caves and you will be able to watch an amazing video of a recent expedition into Wonder Cave.

Peter & Beverly Pickford / Images of Africa

Another view of one of the Cango Caves' impressive halls revealing drape-like cave formations.

Gerhard Dreyer / Images of Africa

Ancient San people, portrayed by this reconstructed family group, probably used the sheltering entrance to the cave for thousands of years. They left a legacy of many rock paintings, now faded.

Immediately outside of the caves you can see the cavernous, colour-stained and partly weathered karstic limestone.

9

Meiringspoort

PROVINCIAL HERITAGE SITE

Superb viewing of the Cape Fold Mountains seen in a dramatically eroded gorge

Meiringspoort is a wonder of both nature and man. The road, recently reconstructed, crosses the Groot River at 25 drifts and provides a wonderful internal spectacle of the Cape Fold Mountains. At Waterfall Drift (seen here) there is a good place to stop and learn more about this geological site.

Between Beaufort West and George, the natural spectacle provided by Meiringspoort, with its extraordinary rock formations, cannot fail to impress and amaze. Travel writer TV Bulpin wrote: 'Meirings Poort remains one of the most romantically lovely of all South Africa's road passes … it is through this splendid natural gateway that the road makes the transition from the Great Karoo to the Little Karoo'.

Location and getting there

- Meiringspoort offers the only tarred road through the Great Swartberg. A section of the N12, the road follows the path of the Groot River gorge. The 13km-long poort starts near Klaarstroom in the north and ends close to De Rust in the south. It crosses the eastern section of the Swartberg Nature Reserve, which is managed by CapeNature, and forms a part of the Cape Floral Region World Heritage Site.
- GPS (Waterfall Drift rest area): 33°24′31″S; 22°33′28″E

Setting the scene

The poort is named after Petrus Johannes Meiring, a local farmer who, in 1854, showed eminent road-builder Andrew Geddes Bain the tortuous route through the gorge. Construction of the road by Adam de Smidt (Bain's son-in-law) was completed in 1857, providing the first proper road through the Great Swartberg into the Great Karoo. However, the route followed the course of the Groot River, fording the river at 25 drifts, and was repeatedly washed away by floods. Over time the road was upgraded but in 1996 yet another destructive flood resulted in the road being closed for months and it was once again reconstructed, with the many fords being reinforced and the road widened.

Start of the poort from the De Rust side, with high peaks of the Swartberg in the distance. The lower hills are part of the Late Proterozoic Cango Caves Group.

Folded, steeply dipping Table Mountain Group quartzitic strata near the southern entrance to the poort.

Regional geology

The entire sedimentary sequence of the **Cape Supergroup** can be seen, as well as the older pre-Cape **Cango Caves Group** near De Rust and younger **Karoo Supergroup** rocks further north. What makes the formations so impressive is the strong deformation of the Cape Orogeny, which created the Cape Fold Belt.

Landscape and landforms

The Great Swartberg is a relict of the ancient pre-African land surface that has eroded over many millions of years. The gorge is another demonstration of the power of river erosion on hard rocks, similar to that seen at Cogmans Kloof (pp.68–71). In the Swartberg, the modest, north-south-trending Groot River has eroded

Seweweekspoort – another geological masterpiece

If you follow Route 62 west from Calitzdorp you will come to Amalienstein where a minor gravel road to Laingsburg (P309) leads northwards to another splendid landscape, perhaps even more impressive than Meiringspoort. Set at river level, a 14km-long pass built between 1859 and 1862, also by Adam de Smidt, follows a river-made cleft through the Swartberg, crossing most of the folded Cape Supergroup, from Table Mountain Group quartzitic sandstone in the south to the valley underlain by Bokkeveld Group shales in the north. The gravel road is good and follows the narrow and convoluted kloof made by the minor Huis River. At the northern end after emerging from the poort, the road splits, one leading to Laingsburg and the N1, the other to Gamkapoort Dam via Bosluis Pass. As with Meiringspoort and Cogmans Kloof, the process of formation of the deeply incised kloof is debatable.

Little-used Seweweekspoort near Ladismith winds its way through the Great Swartberg and shows Cape Fold Mountain geology similar to that seen in Meiringspoort. This photo shows more gentle open folding of Peninsula Formation quartzitic sandstone near the southern end.

The Great Waterfall, above the drift of the same name, is hidden in a side ravine and is not visible from the road. The stream drops 60m into a 9m-deep rock pool and is most impressive when in flood.

through thousands of metres of hard quartzitic sandstone, and yet with little obvious power – it retains its twisting river course, characteristic of mature rivers meandering on flood plains. The down-cutting may result from an earlier drainage, or strong headward erosion from ravines on both sides of the mountain, as seen today on the Dorps River gorge on the northern side of the Swartberg Pass.

Stone walling along part of the old road built by Adam de Smidt in about 1857.

Rock formations

- At the southern side of the poort the somewhat sheared, metamorphosed rocks of the older Cango Caves Group make an appearance.
- This is followed, proceeding northwards, by dipping layers of resistant Table Mountain Group quartzites, split midway by the thin shale of the Cederberg Formation. Most impressive throughout is the strong folding and crumpling of the hard quartzitic strata.
- The northern side of Meiringspoort is composed of alternating layers of Bokkeveld Group shales, siltstones and sandstones, which can be seen along the roadside near Klaarstroom.
- Further north, beyond the poort near Klaarstroom, the whitish ridges are composed of Witteberg Group sandstone, the top group of the Cape sequence.

Things to see and do

- **Drive slowly through the poort:** To appreciate the extraordinary geology you need to drive slowly and make stops at designated parking areas. If the light is right the surrounding mountains will reveal spectacular folding.
- **Stop at the rest area:** In the centre of the poort at Waterfall Drift there is a kiosk with considerable information. Spending time here will give you a worthwhile background on the history of the poort, how the road was constructed and on the natural history of the area.
- **See the Great Waterfall:** A short walk from the rest area will bring you to a waterfall that is well hidden in a side ravine, and which was visited in 1925 by the Prince of Wales. The stream drops some 60m into a 9m-deep mythical 'mermaid' pool and is most impressive after rain.
- **See 'Herrie se Klip':** CJ Langenhoven, famous Afrikaans author-poet and resident of Oudtshoorn, carved the name 'Herrie' into the sandstone alongside the highway at Meiringspoort in 1929 – the name of the imaginary elephant that appeared in his book *Sonde met die Bure* and other stories. This carving (today regarded as undesirable geovandalism) was declared a National Monument in 1971 – now a Provincial Heritage Site.

10

Knysna Heads

GATEWAY TO THE KNYSNA ESTUARY

Rocky quartzite headlands dominate the entrance to the river estuary, once a thriving coastal port

Roger de la Harpe / Africa Imagery

An aerial view of the rugged Knysna Heads, the estuary and Knysna town, with the dissected coastal plain in the middle distance, and in the far distance the Outeniqua Mountains, one of the coastal ranges of the Cape Fold Mountains.

In 1804 George Rex, a well-connected English gentleman, settled along the north shore of the estuary and is considered to have been the founder of the town. Later, the Norwegian Thesen family played a major role in the booming timber and shipping business, utilising the protected harbour provided by Knysna's shallow estuary.

Location and getting there

- The historic coastal town of Knysna lies on the N2 and is situated on the northern side of a shallow lagoon-like estuary. The rocky headlands, called The Heads, are located nearby on the coastline. You can get to the Eastern Head, with its excellent views, by road, but access to the Western Head is restricted as it is part of the private Featherbed Nature Reserve.

- GPS (Eastern Head, upper lookout car park): 34°04′44″S; 23°03′44″E

Setting the scene

The tide enters Knysna's estuary by surging through a narrow channel flanked by two spectacular headlands. In 1818 the first British ship sailed safely through the rocky Knysna Heads into what became a useful harbour. However, although deep, the channel narrows to less than 200m and, with the strong currents, it has always been a treacherous passage in which ships have foundered and lives been lost.

Regional geology

The headlands are composed of strongly jointed, south-dipping strata of hard **Table Mountain Group** quartzite that extends eastwards along the coastline to Cape St Francis, and are part of the **Cape Supergroup** that forms the remarkably resistant Cape Fold Belt. These quartzites are the same rocks that make up the cliffs of Table Mountain.

At the Western Head the quartzite strata dip shallowly to the south but are intensely cut by joints and fractures, which reflect a wide north-south-trending zone of geological weakness, probably fault related.

The Knysna Heads viewed from the Eastern Head rocks, on which a number of ships have foundered, leading to loss of life. These rocks very clearly show the intensity of the north-south fracturing and weakness, probably fault related, which allowed easier erosion by a coastal river gorge between The Heads, geologically similar to Storms River Gorge, but larger.

Landscape and landforms

The Heads are remnants of the extensive coastal plain that is present throughout the southern Cape, and originally part of a wave-cut platform that was later elevated more than 200m above sea level. On their northern side is the wide and shallow estuary of the Knysna River, often referred to as a lagoon. At one time the headlands flanked a narrow river gorge, a feature that is common along the southern Cape's coastal rivers. The river eroded a deep cleft through layers of hard quartzite when the sea was at a lower level but the gorge has since been flooded by the sea. It is speculated that the Western Head, now masked by much younger dune rock, was once an island.

Millwood goldfield

The first gold nugget was found in 1876 in the Karatara River northwest of Knysna and it was only around 1885 that significant alluvial gold was found in other streams such as Jubilee Creek and Millwood Creek, deep in the forest some 20km north of Knysna. There was the inevitable gold rush and the mining village of Millwood soon developed. By 1888 an estimated 700 prospectors were working 1,400 claims and 127kg (about 4,080 ounces) of gold was officially reported between 1887 and 1900. Although most of the production was alluvial gold, there were also underground workings, notably the now refurbished Bendigo Gold Mine, which has a 100m-long adit. But gold recovery was erratic and insufficient to cover costs, so by the early 1890s most of the diggers had moved on, many to the Witwatersrand. Gold mining ceased and considerable heavy mining equipment was abandoned. The gold has its origin in discontinuous, milky white quartz veins related to the Cape Orogeny, found in steeply dipping beds of Table Mountain Group quartzite.

The Materolli Museum at Millwood, in one of the original corrugated iron buildings, has a fascinating collection of mining memorabilia.

Rock formations

Beds of hard, well-jointed Peninsula Formation quartzite of the Table Mountain Group dip at around 20° to the south and are well exposed in the cliffs near sea level. The upper part of the Western Head is covered by much younger dune rock and windblown sand. At the base of cliffs at the Eastern Head and along the seashore at nearby Coney Glen beach a good geological explanation can be seen for the formation of the spectacular passage between the headlands:

- Highly jointed quartzite – there is a very strong set of vertical joints in the quartzite that allows it to break up more easily; in places it has shattered to form zones of brecciation (broken-up rock) and consequent weakness.
- Numerous fractures and thin quartz veins of similar orientation, which are probably fault related and have contributed to a wide zone of structural weakness.
- The result was that intensely jointed and shattered quartzite at this particular site weathered and eroded more easily, and the weakened fracture zone provided an ideal pathway for a young coastal river to cut a gorge.
- North of the quartzite ridge is Knysna estuary, underlain by sediments of the Enon Formation of the Uitenhage Group.

Numerous thin, vertical quartz veins filling fractures in jointed Peninsula quartzite at Coney Glen beach.

Quartzite strata at Coney Glen beach, striking east-west and dipping about 20° to the south. The Western Head is seen in the distance.

Things to see and do

- **The Eastern Head:** From the lookout at the top there are excellent views across the channel to the Western Head and the estuary, as well as the rugged Atlantic coastline, although the best rock exposures are along the lower, sea-level path.
- **Visit Coney Glen beach:** A short walk to this rocky beach provides an interesting geological experience where one can see the shallowly dipping quartzite strata, as well as strong jointing and quartz veining.
- **Sail to the Western Head:** A ferry takes you to the private Featherbed Nature Reserve where you can catch a ride to the top of the headland and go on a short coastal walk to see the geology and views.
- **Visit Millwood goldfield:** Part of the old Bendigo Mine has been refurbished for tourism, and the nearby small Materolli Museum provides historic information on the goldfield. It's also worth making a visit to the relocated Millwood House Museum in Knysna.

11

Robberg Peninsula

PROVINCIAL HERITAGE SITE

An illustration of geological conditions along the southern African coastline after the break-up of Gondwana

Part of Robberg, looking eastwards. The Gap and its beach is in the middle distance. The Island and its linking sand-spit 'tombola' beach is in the far distance on the right, the beach leading up to the (unseen) Witsand dune sand valley.

Robberg, meaning 'mountain of seals' was the name given to the rocky headland by early Dutch mariners. Colonial exploration goes as far back as the 15th century, when the Portuguese named the bay that's partly enclosed by the Robberg headland as Bahia Formosa (Beautiful Bay).

Location and getting there

- Robberg lies some 5km to the south of Plettenberg Bay town, and you get there by diverting off the N2 and following Robberg Road southwards. The peninsula forms the 245ha Robberg Nature Reserve administered by CapeNature, and there is a charge to enter.
- GPS (car park at entrance): 34°06′02″S; 23°22′39″E

Setting the scene

The first white settlers arrived at Formosa Bay in the 1760s; at that time the Khoikhoi people lived along the coast, the probable descendants of 'Strandloper' people who had lived there for many thousands of years. Robberg Peninsula is an unusual promontory, a declared National Monument (now Provincial Heritage Site), and provides excellent lessons in geology, telling of a sedimentary basin that developed as the new southern African coastline was formed after the break-up of Gondwana. It is also the site of the famous Nelson Bay Cave. True to its name, it has a thriving colony of Cape fur seals and a large diversity of marine bird life.

A walk down the steep slope on the south side brings you to the archaeologically excavated Nelson Bay Cave, which was formed in angular quartzite breccia of the Robberg Formation by erosion from the sea. It was occupied by early humans as far back as 120,000 years ago.

Regional geology

The main rock groups are a resistant basement of the **Cape Supergroup's Table Mountain Group** quartzite and of the **Robberg Formation**, which is related to the **Uitenhage Group**. The latter was deposited in fault-created intermontane valleys and offshore basins that formed during the southern break-up of Gondwana around 120 million years ago. Large amounts of coarse sediment came from strong erosion of the high Cape Fold Mountains.

A package of thick Robberg conglomerate makes up most of the cliff at The Gap and is topped by a layer of reddish sandstone. Inset: The conglomerate consists of well-rounded pebbles and cobbles of hard quartzite, and is probably related in time to the Enon conglomerates of the Gamtoos Valley (see pp.102–105).

Landscape and landforms

Robberg is almost 4km long and up to 750m wide, and rises to 148m above sea level. It is being actively eroded by the sea and wind because of its exposed coastal position. It narrows down to less than 200m at a fracture zone called The Gap, which is no more than 20m above sea level, and where it could be breached in the distant future.

Rugged Robberg Peninsula at The Gap, as one looks eastwards, and the best place to see the geology. The hiking paths can be seen on both sides. The Indian Ocean is on the right and Formosa Bay on the left.

Rock formations

Three distinct geological formations of different ages make up Robberg Peninsula, as shown on the annotated photo (opposite, bottom):

- Table Mountain Group: The quartzite (recrystallised sandstone) is part of the Peninsula Formation. It forms hard south-dipping beds that are well exposed close to sea level, and makes the rigid basement upon which the younger Robberg Formation was later deposited.
- Robberg Formation: This is the dominant formation and comprises various sandstones, conglomerates and scree breccia. It is up to 80m thick and was deposited around 120 million years ago. From the base upwards it consists of very irregular deposits of broken rock fragments, now cemented to form hard breccia; on top of which are impressively thick conglomerate beds that formed on pebbly beaches; and this was followed by shallow marine sandstones.
- Quaternary dune rock: This is up to 30m thick and is mostly restricted to the highest part of Robberg. These rocks are also well exposed on The Island, an isolated outcrop that lies to the south. Young, unconsolidated windblown beach sand covers part of Robberg today, for example at the Witsand valley.

History of the Robberg Peninsula

Robberg provides a small but fascinating geological 'window' into the development of a sedimentary basin as the new southern African coastline was being formed, following the break-up of Gondwana. It probably represents part of the northern shoreline of the large, offshore Pletmos Basin that lies submerged on the Agulhas Shelf, and its history can be summarised as follows:

- Before the Robberg sediments were deposited there was a rugged, scarp-like ridge made of hard, south-dipping Table Mountain quartzite, perhaps like the Knysna Heads are today.
- Strong coastal erosion of the ridge first formed deposits of unsorted rock debris (scree) that accumulated close to its source area. The scree hardened to form a lower breccia of angular quartzite fragments cemented within a finer-grained matrix.
- Active regional faulting caused the basin to subside below sea level and strong wave action produced pebbly beach deposits that are now preserved as layers of well-rounded quartzite pebble conglomerate.
- Renewed subsidence once again led to the sea flooding in, but in more subdued conditions, and this added a complex sequence of tidal and channel sands on top of the conglomerates, and formed the upper sandstone layer.
- Later, during the Cenozoic, renewed erosion and uplift produced a regional land surface that survives near the top of Robberg, upon which windblown sand dunes were later formed, to become cemented dune rock.

Things to see and do

- **Do the Robberg hikes:** There are three popular circular hikes across Robberg, of increasing distance and difficulty, and all provide excellent geological exposures. The longest is to The Point (over 9km round trip) and the shortest is the easy walk to The Gap, which takes less than an hour. Admire the superb views and animal life on the way.
- **Visit Nelson Bay Cave:** The cave has been excavated by archaeologists and tells a fascinating story of early human occupation from 120,000 years ago, and later by San hunter-gatherers and Khoikhoi people. It is supported by a good interpretation facility. The cave is a large overhang formed by wave erosion at the unconformity between quartzite and breccia. Ghwano Cave east of The Gap is similar but is not accessible to visitors.

12 Storms River Gorge

PART OF GARDEN ROUTE NATIONAL PARK

A deep and narrow river chasm that has been cut into the elevated coastal plain

Storms River Mouth from the lookout point, as one faces west. This is the ocean side of a zone where long-past geological forces have cracked and broken the steeply dipping Table Mountain Group quartzite strata at about 90° to their general east-west trend to form a narrow zone of fracturing that continues inland for more than 3km. This is essentially the same process of fracturing and erosion (only narrower here) that took place at Knysna Heads.

The region south of the Tsitsikamma Mountains is well known for its rugged coastline, indigenous coastal forest and the famous 50km-long Otter Trail. Here the incredible Storms River Gorge – one of several deeply incised valleys that cut through the coastal plain – enters the sea at the picturesque resort of Storms River Mouth.

Location and getting there

- The bridge on the N2 over Storms River Gorge is some 65km east of Plettenberg Bay. The popular resort at Storms River Mouth is in the Tsitsikamma section of the Garden Route National Park, a narrow conservation strip that extends 80km along the coast and includes a 5km-wide offshore marine reserve. Divert off the N2 and follow the road to Storms River Mouth for 8.5km.

- GPS (suspension bridge): 34°01′10″S; 23°54′13″E

Setting the scene

East of Plettenberg Bay the N2 crosses a remarkably flat plain that is cut at intervals by deep river gorges, now spanned by modern, high-level bridges. It has not always been so, as the old road was cut through almost impenetrable forests and followed steep winding passes though the Grootrivier, Bloukrans and Storms River valleys. In the early 1880s famous road-builder Thomas Bain built the first passes, said to follow ancient elephant trails.

Storms River Gorge, as one looks south, here about 120m deep and less than 10m wide, seen from the bridge on the N2. Notice the flat coastal plain in the background.

The Paul Sauer Bridge, which opened in 1956, spans some 100m across the Storms River Gorge. It is firmly anchored in Table Mountain Group quartzite and was constructed in a cantilever style from both sides.

Regional geology

The elevated coastal plain is mostly underlain by strongly folded formations of the **Table Mountain Group** (exposed along the rugged coastline), part of the **Cape Supergroup**. These rocks were compressed and deformed during the formation of the Cape Fold Belt, resulting in a more or less east–west structural grain.

Landscape and landforms

The narrow, elevated, remarkably flat coastal plain, backed by the rugged Tsitsikamma Mountains, tilts gently towards the sea. Now uplifted to an average elevation of over 200m, it originated as a wave-cut platform that formed at sea level. It is cut by several deep river gorges that originated as zones of weakness exploited by short coastal rivers (see text box below).

Closer to Storms River Mouth the north-south-oriented river gorge is impressively straight over a distance of about 3km, and almost impossibly narrow. It cuts the structural grain at about 90° and follows a narrow zone of fracturing that weakened the rocks and allowed the youthful Storms River to cut down effectively through hard quartzite, keeping pace with uplift.

Rock formations

The geological map shows that from the Tsitsikamma Mountains southwards to the sea, all the quartzite formations of the Table Mountain Group are present, as well as the shale-rich Bokkeveld Group. While this is not obvious on the ground, the rocks are tightly deformed into a steeply dipping, east-west-trending syncline in the north and an anticline in the south, causing repetition of strata.

● Hard, resistant Peninsula Formation quartzite, as seen in the Tsitsikamma Mountains, is repeated in the straight 3km-long

The southern Cape coastal plain

The mouth of the Storms River is on the left, as seen from the resort. The high ground in the distance is the now-elevated coastal plain, terminating in steep sea cliffs. Steeply dipping Table Mountain quartzite and siltstone are in the foreground.

A remarkable feature of much of the southern Cape's coastal landscape is the elevated plain that lies between the sea and the mountains. In the Storms River area it is up to 7km wide and is known as the Aberdeen Plain. It slopes very gently seaward and ends in awesome coastal cliffs, and is dramatically and deeply incised in places by short coastal rivers. This is a wave-cut platform that was formed by the pounding action of waves from the open ocean eroding the coastline during a marine transgression (sea-level rise). This took place in the Miocene when the shape of the coastline was much the same as it is today. The unrelenting battering that it received produced a very gently sloping surface, mostly across incredibly hard Table Mountain quartzite. Later vertical earth movement along the southern coast elevated the wave-cut platform to its present height, now between 200m and 230m above sea level, and produced the spectacular coastal scenery.

gorge running inland at Storms River Mouth.

- A bush-covered cleft in the hillside adjacent to the suspension bridge marks the position of the Cederberg shale unit separating the Peninsula and Nardouw quartzites.
- And to the south, all along the rocky coast, are brownish-weathering, alternating Nardouw quartzites and siltstones of the upper part of the Table Mountain Group.

All these beds dip very steeply to the south. Near the parking area there are very noticeable small-scale folds, the axes of which plunge shallowly to the east.

At the suspension bridge across the river, as one looks eastwards, the steeply south-dipping quartzite strata can be seen very clearly.

Close to the resort area several elongated small-scale folds can be seen; here, one is looking down the fold axis of a superb small synclinal structure in quartzite.

In places along the sea front rocks there are irregular white quartz veins that cut across or run parallel to the strata; there are rumours of gold occurring here.

Things to see and do

- **Kayak up the river:** This is the best way to experience the drama of the gorge and see the geology; the walls are so high and narrow that at times the sky is almost lost.
- **Go tubing down the river:** Starting at Storms River village this takes you floating, clambering and jumping down-river and you will get an excellent view of the passing geology.
- **See Paul Sauer Bridge:** On the N2 the bridge crosses the Storms River about 6km from the coast. The exceptionally narrow gorge bottoms some 120m below road level and is not more than 10m wide.
- **Hike the first part of the Otter Trail:** This takes you on a sea-side scramble where you can see remarkable coastal geology, a most impressive sea cave, and a splendid waterfall no more than 2.5km along the trail.
- **Look for the gold:** The occurrence of gold (*goudgate*) is marked on the geological map near the start of the Otter Trail. Nearby there are numerous white quartz and brown carbonate veins in quartzite and phyllite, some of which apparently contain pyrite and traces of gold.

13 Gamtoos Conglomerates

QUEEN VICTORIA REMEMBERED

Amazingly thick beds of conglomerate reveal a time of very large-scale and rapid erosion during the break-up of Gondwana

At Andrieskraal, about 15km west of Patensie, there are amazingly high cliffs made entirely of pebble-to-boulder-sized conglomerate of the Enon Formation, which stand almost vertically without obvious support. Inset: One cliff exposure displays a remarkable likeness to the profile of the elderly Queen Victoria.

Picturesque Gamtoos Valley is one of the Eastern Cape's major agricultural districts, noted for its citrus farms. Geologically it holds remnants of younger sedimentary formations deposited in valleys during Late Jurassic to Early Cretaceous times, when large amounts of sediment derived from the eroding Cape Fold Mountains were deposited along the coastal margin of the new African continent.

Location and getting there

- The best conglomerate exposures are at Andrieskraal near the head of the Gamtoos Valley. From Hankey (on the R330) you need to get onto the road to Patensie (R331) and then travel about 15km further west. Or you can approach from the west on the R332 via Baviaanskloof (note: 4x4s only).
- GPS (bridge near cliffs): 33°45′28″S; 24°41′21″E

Setting the scene

There are excellent exposures of incredibly thick conglomerate formation along the riverside in the Andrieskraal area, where spectacular white and red cliffs rise more than 100m above the Gamtoos flood plain. One such exposure has famously been likened to a profile of the elderly Queen Victoria. Conglomerate exposures are also seen in other high cliffs in the Hankey area. These exposures lie on the road to the Baviaanskloof Wilderness Area, part of the Cape Floral Region World Heritage Site.

Regional geology

Spectacular conglomerates that make up the **Enon Formation**, the lowermost division of the **Uitenhage Group**, extend discontinuously from west of Oudtshoorn to Algoa Bay in the east, and are part of several disconnected, fault-controlled sedimentary basins from the Cretaceous. One of these is the 40km-long Gamtoos Basin, sandwiched between mountains of the Cape Fold Belt in the southeastern Cape. The same formations also exist offshore on the continental shelf, where they form the extensive sediments of the Outeniqua Basin.

Looking across the fertile Gamtoos Valley flood plain, with the Late Proterozoic Gamtoos Group forming the bush-covered hills in the middle distance. In the far distance is the 1,768m-high Cockscomb Mountain, part of the Cape Fold Mountains.

Landscape and landforms

The present-day Gamtoos Valley forms an alluvial flood plain, bordered by hills made largely of Cretaceous-age sedimentary conglomerates, and sandwiched between Cape Fold Mountains. This may not be too different from the landscape at the time of deposition of the coarse sediments, but the mountains would have been higher and steeper. Today the compacted thickness of exposed conglomerates stands up well to erosion and in places cavities have formed along steep faces, particularly within the reddish conglomerate zone.

Rock formations

The sequence of Enon Formation conglomerate at Andrieskraal is well over 100m thick. There are two distinct zones:

- A lower, very poorly bedded 'white Enon', in which the whitish conglomerate is clearly exposed along the road and in the road cutting, where it is made of smooth, tightly packed boulders, cobbles and pebbles that are poorly sorted. The clasts are generally well rounded but vary considerably in size, from less than 5cm to over 25cm, and are almost exclusively of hard, resistant Table Mountain quartzite. The cementing matrix consists of sand and silt, and minor limonite (iron oxide) that coats some of the clasts. Within this sequence there are narrow layers or lens-like bodies, of limited extent, of whitish sandstone and siltstone; in places they are impregnated with limonite.
- An upper 'red Enon' that is much better bedded and contains sandstone layers that are reddish, with more angular pebbles and a higher content of iron oxide.

Steep hills west of Patensie, made of Enon Formation conglomerate of both the lower 'white' variety and the overlying more cavernous 'red' variety. Top inset: The 'red' variety in which the pebbles are more angular and have an abundant coating of limonite (rusty iron oxide). Lower inset: The mixed sizes of rounded quartzite pebbles and cobbles that can be seen alongside the road at Andrieskraal.

Geological history

Around 120 million years ago, as South America pulled away from Africa during the break-up of western Gondwana, a major zone of transform faulting called the Agulhas-Falkland Fracture Zone developed immediately south of what is today's southern Cape coast.

This large-scale break-up of Gondwana resulted in major faulting and rifting within the ancestral – and, at that stage, much higher – Cape Fold Mountains, following the east–west structural grain, and leading to the development of a series of fault-bounded, intermontane basins.

Strong erosion of the high Cape Fold Mountains resulted in fast-flowing, gravel-filled streams and rivers. Considerable coarse sedimentary detritus, almost exclusively from resistant Table Mountain quartzitic sandstones, was produced.

This poorly sorted sediment, consisting of boulders, cobbles, pebbles and sand, was deposited on land as mountain-side alluvial fans, while finer sand and silt washed downstream to flood plains.

Sediment accumulated at the base of active fault scarps and subsidence of the faulted basins kept pace with sediment being deposited. The result was exceptionally thick but localised deposits that are now the Enon conglomerates.

Things to see and do

- **See the Andrieskraal conglomerates:** Spectacular exposures of white Enon conglomerates can be seen at road level, and it is worth parking nearby to examine these rocks more closely. Look upwards to see the red Enon conglomerates and numerous cavities.
- **Look for Queens Victoria's profile:** This is close to the bridge at Andrieskraal and is best seen by approaching the cliffs from the west.
- **Drive to Kouga Dam:** At the head of the Gamtoos Valley is the long, sinuous and narrow Kouga Dam, the water held back by a double-arch dam wall built in 1967. The dam is hemmed in by steeply dipping strata of the Cape Fold Mountains.
- **Visit the grave of Sarah (Saartjie) Baartman:** This site is of considerable cultural heritage importance. Located on high ground overlooking the Gamtoos Valley at Hankey, it provides an excellent overview of the geological formations discussed here.

The inscribed memorial to Sarah Baartman (c.1789–1815), a Khoikhoi slave who was taken to Europe as a human curiosity. She died in France, and her remains were repatriated to her homeland and buried at Hankey in 2002.

14 Valley of Desolation

PROVINCIAL HERITAGE SITE

Magnificent views of the Great Karoo from a spectacularly eroded dolerite sill

Some of the impressive dolerite columns at the Valley of Desolation, as one looks southwest across the Great Karoo. These columns formed as a result of vertical jointing, caused by cooling and contraction of a thick Karoo dolerite sill intruded into the thick existing Karoo Basin stratigraphy.

This geosite was declared a National Monument in 1939 because of its dramatic scenery and geological significance, and is now a Provincial Heritage Site. There are panoramic views over Graaff-Reinet and the Sundays River Valley, across the Plains of Camdeboo to the south and to the Sneeuberg Escarpment in the north.

Location and getting there

- The Valley of Desolation lies within SANParks Camdeboo National Park, which almost completely surrounds the town of Graaff-Reinet. Superb mountain-top viewing sites are about 14km by tarred road from Graaff-Reinet: take the Murraysburg road (R63) for about 3km, turn left into the park and follow the signposted mountain road.
- GPS (parking area on top): 32°15′51″S; 24°29′36″E

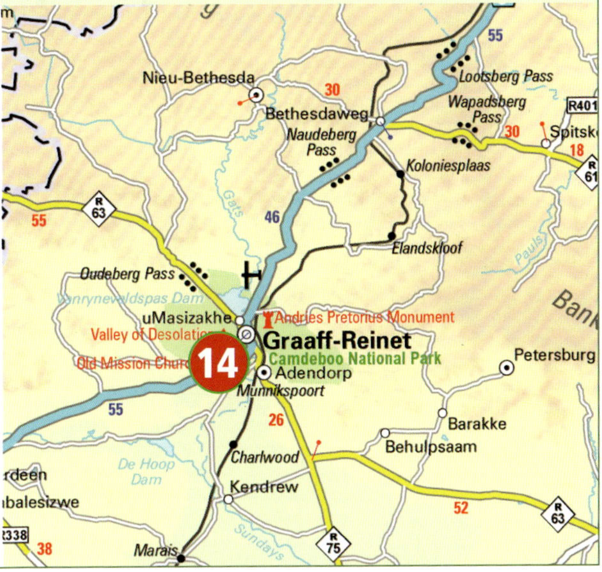

Setting the scene

The spectacular landscape provided by the Valley of Desolation has long been a tourist attraction, and it has also been called the Cathedral of the Mountains. The name originally referred to a narrow eroded gorge or 'valley' on the edge of the dolerite scarp but now applies to the entire massif. The Karoo Nature Reserve was established here in 1979 and expanded in size and proclaimed a National Park in 2005.

Regional geology

Part of the **Karoo Supergroup**, the formations here are comprised of flat-lying sedimentary rocks of the **Beaufort Group**, making them about 250 million years in age. At a later stage, around 180 million years ago, these horizontally layered sedimentary rocks were intruded by **Karoo dolerite** magma, which solidified and slowly cooled to form thick and widespread sills, and dykes.

The tops of some large dolerite columns, which can be several metres across. Notice the joints and fractures that formed as a result of cooling and slow contraction of the dolerite sill. Inset: A close-up view of similar fresh dolerite sill seen in a road cutting near Nqweba Dam (formerly Van Ryneveld Pass Dam).

From the western end of the Crag Lizard Trail there's a good view of the collapsing dolerite columns and of Spandau Kop. This pointed hill, seen here at a distance of about 3km, is an excellent example of a butte.

Landscape and landforms

The area lies close to the dissected Sneeuberg range of the southern Great Escarpment, which is formed here by numerous and extensive intrusions of resistant dolerite that have partly protected the underlying mudstones and sandstones from erosion. The Valley of Desolation is a remnant of that dolerite escarpment, which has now largely retreated further north. The nearby iconic Spandau Kop (1,316m) is an isolated butte capped by a probable remnant of the same dolerite sill.

The peculiar column-like wall of the dolerite sill is called 'columnar jointing', a feature that is often seen in Karoo sills (see Platberg, pp.150–153). After intrusion deep underground, the sill cooled and solidified very slowly; and as it cooled it contracted slightly and developed an internal network of shrinkage cracks or joints. Predominantly vertical, these joints produce large, roughly polygonal to squarish vertical structures, as well as horizontal joints. All these joints are surfaces of weakness that, on exposure to the elements over time, are selectively exploited by agents of weathering and erosion. The result is the impressive stacked columns and vertical walls of eroded dolerite, which are estimated to be about 35m high, although in places they have collapsed into blocky debris that now litters the floor of the valley.

Rock formations

The thick dolerite sill extends for many kilometres around, forming the higher ground.

The flat-topped mountain at the Valley of Desolation, a remnant of this once more extensive sill, has baked the underlying Karoo sediments. Lower down, near the Nqweba Dam wall, another thick dolerite sill makes an appearance, perhaps originally part of the same intrusion. There are also several vertical dolerite dykes in the area. Close-up, the fresh dolerite has a medium-grained crystalline texture, and is dark grey when fresh and dark rusty brown when weathered. The dolerite came from the same magma source that produced the Drakensberg basalt flows (see Sani Pass, pp.123–127) but the molten magma was also forced into beds of existing sedimentary rock deep underground, where it solidified. Subsequently, over millions of years, major continental uplift took place followed by vast denudation that removed the softer sedimentary rocks and exposed the harder dolerite sill to the elements.

A fossil trip to Nieu-Bethesda

About 50km north of Graaff-Reinet, and a bit off the N9, lies the quaint village of Nieu-Bethesda, well known for the Owl House, now a shrine to the eccentricity of its former owner, the late Helen Martins. But don't miss the excellent Kitching Fossil Exploration Centre across the road, named for internationally renowned palaeontologist Dr James Kitching, who grew up in Nieu-Bethesda and was one of South Africa's best 'fossil-finders'. This small museum tells the fascinating story of life and death on the original Karoo flood plain during the Permian (see also Fraserburg Fossil Surface, pp.307–311) only a few million years before the Earth suffered a mass extinction around 252 million years ago, when over 90% of life was wiped out. The region around Nieu-Bethesda is renowned for its Karoo reptile fossils and it was from here that the local Rubidge family recorded some of the earliest Karoo fossil finds, and from where the famous Dr Robert Broom described many Karoo fossils. At the museum you will see the gorgonopsians ('terrible eyes') which preyed on plant-eating dicynodonts ('two dog teeth'). Life-like reconstructions are on display as well as fossil bones and recreated scenes of the Karoo palaeo-environment. You can walk to the usually dry bed of the Gats River where you will see numerous fossil bones still embedded in their enclosing rock. Here, before 252 million years ago life prospered on the vast alluvial flood plain of the original Karoo, where, as animals died, their remains became embedded in muddy alluvium to be preserved as fossils.

Things to see and do

- **See the Valley of Desolation:** The Mountain Drive takes you through impressive Karoo scenery. Stop off at the Toposcope to look over Graaff-Reinet nestling in a loop of the Sundays River, then go to the Wall lookout and see the Great Karoo from an elevation of almost 1,400m.
- **Walk the Crag Lizard Trail:** Taking less than 1 hour, this highly recommended circular trail starts from the Wall and takes you to the western end of the mountain and back to the car park. You will get to a number of spectacular viewing points and also get close to the dolerite outcrops.
- **Visit the Kitching Fossil Exploration Centre:** Located in the quaint village of Nieu-Bethesda about 50km north of Graaff-Reinet, this small but important museum tells the story of life and death on an early Karoo flood plain around 250 million years ago. You will be able to see real fossil bones still embedded in rock.

15 Hole in the Wall

ICONIC LANDFORM OF THE WILD COAST

A spectacular coastal landmark, an uncommon stack topped by Karoo dolerite

Claudio Velasquez Rojas / Homebrew Films

The Hole in the Wall islet seen offshore from the air. The rock mass, capped by a thick and columnar-jointed Karoo dolerite sill, is an unusual sea-stack and the hole is oddly lined up with the Mpako River in the background. Notice how the rocky mass has eroded more above and around the hole, and will one day collapse.

Not far south of Coffee Bay there is a fascinating off-shore islet known as Hole in the Wall, or called by its Xhosa name, eziKhaleni, meaning the 'place of the sound'. This is one of South Africa's rare sea-stacks, a large, free-standing rocky outcrop through which a spectacular portal has been bored by countless years of pounding waves.

Location and getting there

- Hole in the Wall lies about 8km from the coastal village of Coffee Bay on South Africa's Wild Coast, at the mouth of the small Mpako River. The only practical way to get there is off the N2 at Viedgesville, 19km southwest of Mthatha, and then follow the tarred road via Mqanduli to Coffee Bay about 90km away. There is ample seaside accommodation in this popular resort area.
- GPS (riverside beach): 32°02′25″S; 29°06′31″E

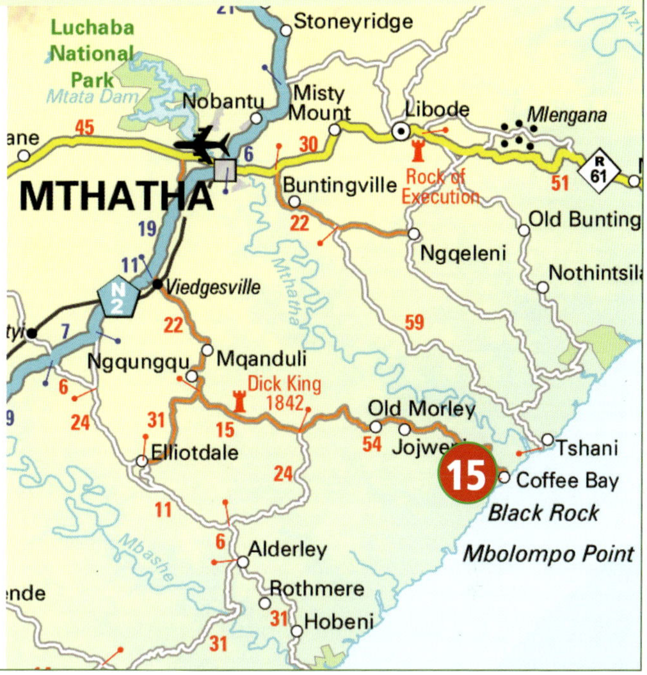

Setting the scene

The Wild Coast along the Eastern Cape's shoreline remains one of South Africa's least visited and less explored regions, where the traditional pace of life remains relatively unaffected by modern life. It is a land of spectacular coastal scenery, rivalling that of the southern Cape, although access is often difficult due to nonexistent or poor roads.

The landmark coastal stack was named in 1823 by Captain Alexander Vidal while he was engaged in surveying the coastline from his Royal Navy ship, HMS Barracouta.

Regional geology

The **Ecca Group** of the **Karoo Supergroup**, comprising dark shale, mudstone and subordinate sandstone layers, forms the underlying geology of this region but, other than along the coastline, exposures are generally poor. The formations are either flat-lying or show a gentle dip to the north. Intrusions of **Karoo dolerite** form numerous dykes, sills and inclined sheets, which present spectacular coastal landscapes. On a larger scale, the generally straight northeast trend of the coastline is a relict of much earlier coastal rifting that is associated with the break-up of Gondwana.

Landscape and landforms

A stack is defined as 'an isolated mass of rock near a coastline, detached from the main mass usually by marine erosion (especially wave action) and rising steeply from the surrounding sea'. This is a perfect description for the Hole in the Wall feature. The stack is oriented northeast-southwest, is about 250m long, up to 55m wide and more than 40m high. It was once part of the coastline, which has been mostly eroded

Walter Knirr / Images of Africa

A closer-up view of the actual arch-like Hole in the Wall. It is about 15m wide and up to 8m high and has been shaped by pounding sea eroding the Ecca sediments. The disrupted lower dolerite sill can clearly be seen above the hole.

back, leaving the stack isolated. Where it is exposed to the open sea on its eastern side, it is considerably eroded away above the hole. The famous arch-like hole is estimated to be roughly 8m high and more than 15m wide at sea level, the softer sedimentary strata beneath the lower dolerite sill having been eroded and excavated by the ever-present wave action. The hole is aligned with the river that drains the hinterland but the reason for its position roughly opposite the river mouth is not clearly understood.

Rock formations

There are two types of rock formation: the original Ecca Group sedimentary rocks are interlayered with two sills of Karoo dolerite – both dipping north at about 5–10°. At the base there is reddish-weathering sandstone and/or siltstone; above this is an estimated 2–3m-thick dolerite sill and above the dolerite is blackish-weathering shale. The sedimentary rocks were almost certainly hardened (indurated) by heat, in particular

Roger Bosch / Media Club South Africa

A large wave crashes through the hole – note the fisherman on the right for a sense of the scale of the hole.

Roger de la Harpe / Africa Imagery

Another wave-made hole known as Baby Hole has formed on the nearby coastline.

The legend of the Abamlambo – 'people of the water'

Xhosa mythology has a very different and much more romantic explanation of how the hole was formed. According to local legend the river was once a landlocked lagoon blocked from the sea by a massive cliff, and living in the sea there were mythological characters called the Abamlambo ('sea people') who had supple, flipper-like feet and hands. One of the sea people noticed a beautiful young girl who lived in a village near the river and they fell in love, but her father forbade the love match. One night the girl slipped away to meet her lover and told him about her father's disapproval, and he devised a solution. At high tide the Abamlambo brought a mighty fish that battered its way through the base of the cliff and the sea flooded in through the breach. Singing and shouting, the sea people rode in on the wave, the girl rushed to meet her lover who swept her up, and as the wave retreated the Xhosa maiden went with the Abamlambo back through the hole, and was never seen again. And it is said that, at certain times of the year, the music and singing of the 'people of the water' can still be heard.

from the thick upper sill that caps the stack – which was not volcanic but formed from extremely hot intrusive magma at probably around 1,000°C. This upper sill is considerably thicker than the lower one, and continues onto the land as a large, cliff-like mass about 100m to the north, and also to the south. Both sills show pronounced vertical jointing, and the lower sill shows what appears to be a structural disruption immediately above the hole. Resistant dolerite boulders and pebbles derived from the sills litter the beach.

Roger Bosch / Media Club South Africa

The Hole in the Wall massif as seen from the riverside beach. The rounded boulders in the background consist of eroded dolerite from the sills.

Things to see and do

- **View and visit the stack:** There are many magnificent viewing points from various locations but it is best to view this natural geological wonder from the safety of the beach. At low tide it is possible to wade to the inner rock ledge close to the hole.

CAUTION: Do not climb the stack. It is surrounded by almost vertical, precipitous cliffs and climbing by amateurs is not recommended, nor is swimming through the hole advised. There have been several unfortunate incidents in the past, including deaths.

CENTRAL KWAZULU-NATAL AND FREE STATE

OVERVIEW OF LANDSCAPES AND GEOLOGY

In this region the geology ranges in age from Middle Proterozoic rocks of the Natal Metamorphic Province to the Jurassic, when the great Karoo igneous event took place. Geosites 16–23 are shown on the national maps (pp.12–13), and all except Geosites 16 and 23 (which fall outside of the illustrated area) on the regional maps (pp.116–117). This region is dominated by the Great Escarpment, created by Drakensberg lavas that are still being eroded back from the coast. Here the mountains are at their highest, over 3,000m in places, leading onto the dissected mountain plateau of the Lesotho Highlands. Below the lava is the Little Berg, characterised by Clarens sandstone, and iconic at Golden Gate. In the east, granites and gneisses of the Natal Metamorphic Province, partly protected by hard horizontal Natal sandstone, form a landscape deeply incised by active coastal rivers.

The major geological formations of the region are listed below:

■ Natal Metamorphic Province (Middle Proterozoic)

The rocks form several discontinuous, fault-bounded inliers from Port Edward in the south to near Eshowe in the north, and along KwaZulu-Natal's South Coast. The various granites and gneisses were formed 1,200–1,000 million years ago during the major Namaqua mountain-building event, which created a wide and structurally complex belt along the southern margin of the Kaapvaal Craton. They are part of similar formations present in the Northern Cape, more than 1,000km to the east (see also p.287). Good examples of these igneous and complex metamorphic rocks are seen in the Oribi Gorge (**Geosite 16**) and the Valley of a Thousand Hills (**Geosite 20**), which were once covered by a protective layer of hard Natal or Msikaba sandstone.

■ Natal Group (Ordovician) and Msikaba Formation (Devonian)

Sporadic outcrops of the **Natal Group** form most of the flat-topped hills and cliffs in the northern part of the Coastal Hinterland of eastern KwaZulu-Natal. This hard, horizontal sandstone is well known from the area around Marianhill and Inchanga, and forms the spectacular cliffs at KwaZulu-Natal's Table Mountain (**Geosite 20**) north of Camperdown. This sequence is up to 600m thick and is dominated by reddish to maroon quartz-rich sandstone that was deposited by rivers on a vast alluvial plain around 490 million years ago. These rocks are correlated in time with the **Table Mountain Group** of the **Cape Supergroup**. Further south (not shown on regional maps), the thick and very similar-looking **Msikaba Formation** is present; also composed of coarse-grained, quartz-rich sandstone, often gritty to pebbly, it was deposited under shallow marine conditions (Oribi Gorge, **Geosite 16**). It was believed that the Msikaba rocks were part of the Natal Group but recent work has changed this thinking. The presence of Devonian fossils shows that the Msikaba Formation is similar in age to the **Witteberg Group** of the Cape Supergroup, and therefore millions of years younger than the Natal sandstone found further north.

Karoo Supergroup
(Carboniferous to Jurassic)

Most of central KwaZulu-Natal is underlain by generally horizontal Karoo strata, which dominate the landscape, from glacial **Dwyka tillite** at the base to its volcanic top layer of **Drakensberg Group** lavas. In between are the thick and varied sedimentary deposits of the **Ecca, Beaufort** and **Stormberg Groups**. Geological sites in this part of the Karoo sequence are mostly associated with Clarens sandstone and Drakensberg lavas. The exception is Howick Falls, which is within Ecca shales.

The Stormberg Group (informal name) (Triassic to Jurassic) comprises the **Clarens**, **Elliot** and **Molteno Formations** – the upper sedimentary formations of the Karoo Basin, where conditions were becoming progressively more arid. The Clarens sandstone, over 100m thick, reflects the desert conditions that existed on Gondwana as far north as the Limpopo, where very similar windblown (aeolian) sandstone beds can be seen at Mapungubwe (pp.224–229). Below this desert sandstone are the Elliot and Molteno sedimentary strata, which represent much more hospitable times, when life prospered at stages on alluvial flood plains and where early dinosaurs made their nests, such as at Golden Gate (**Geosite 21**). Clarens sandstone is also well displayed at Platberg (**Geosite 22**). Over time the thick accumulation of Karoo sediments became hardened into rock. Around 183 million years ago vast eruptions of molten basalt lava gave rise to the thick **Drakensberg Group**. Over a geologically short period of a few million years, countless thin lava

The Drakensberg range, much of which has been declared a cultural and natural World Heritage Site, is one of South Africa's prime recreational mountain regions. Rising to over 3,000m in places, its ramparts form the precipitous border with neighbouring Lesotho. This photograph shows the mountains in the Sani Pass area.

A remnant pinnacle of weathered lava in the Drakensberg, seen on the Sani Pass.

flows covered the landscape across much of southern Africa. This igneous activity took place when a hot mantle plume developed under the northeastern part of southern Africa; the crust began to crack along deep, mantle-tapping fractures that would later lead to continental rifting and the break-up of Gondwana. Today it is the Drakensberg that reminds us of the enormous lava province that once existed, where the dramatic mountain scenery defines the precipitous break between the dissected Coastal Hinterland of KwaZulu-Natal and the ancient erosional plateau of the Lesotho Highlands, as seen at Sani Pass (**Geosite 17**) and at the Amphitheatre (**Geosite 18**).

Karoo Dolerites
(Jurassic)

Much of the scenery in KwaZulu-Natal's Karoo formations is related to intrusions of dolerite, part of the igneous event that produced the immense outpourings of Drakensberg basalt lava (see above, and also the description in Southern and Eastern Cape, p.73). Under considerable pressure and at an initial temperature of up to 1,200°C, red-hot magma was squeezed upwards into fractures many kilometres in length, and along horizontal bedding planes deep within the now lithified Karoo sedimentary rocks. It was these fractures that also provided the extensive magma conduits for the overlying Drakensberg lavas, which have the same chemical composition and magma source. A look at the regional geological map reveals the extent of the area underlain by Karoo dolerite, and seen as the thick sills at Howick Falls (**Geosite 19**) and the rock that caps Platberg (**Geosite 22**).

Figure 16: Landscape map of central KwaZulu-Natal (KZN) and part of the eastern Free State region. This is dominated by three geomorphic provinces: the high plateau of the Lesotho Highlands, including the Drakensberg Escarpment, and the Highveld, both coloured brown; the Midlands in tones of orange and yellow; and the lower Coastal Hinterland in green and mostly light blue-grey.

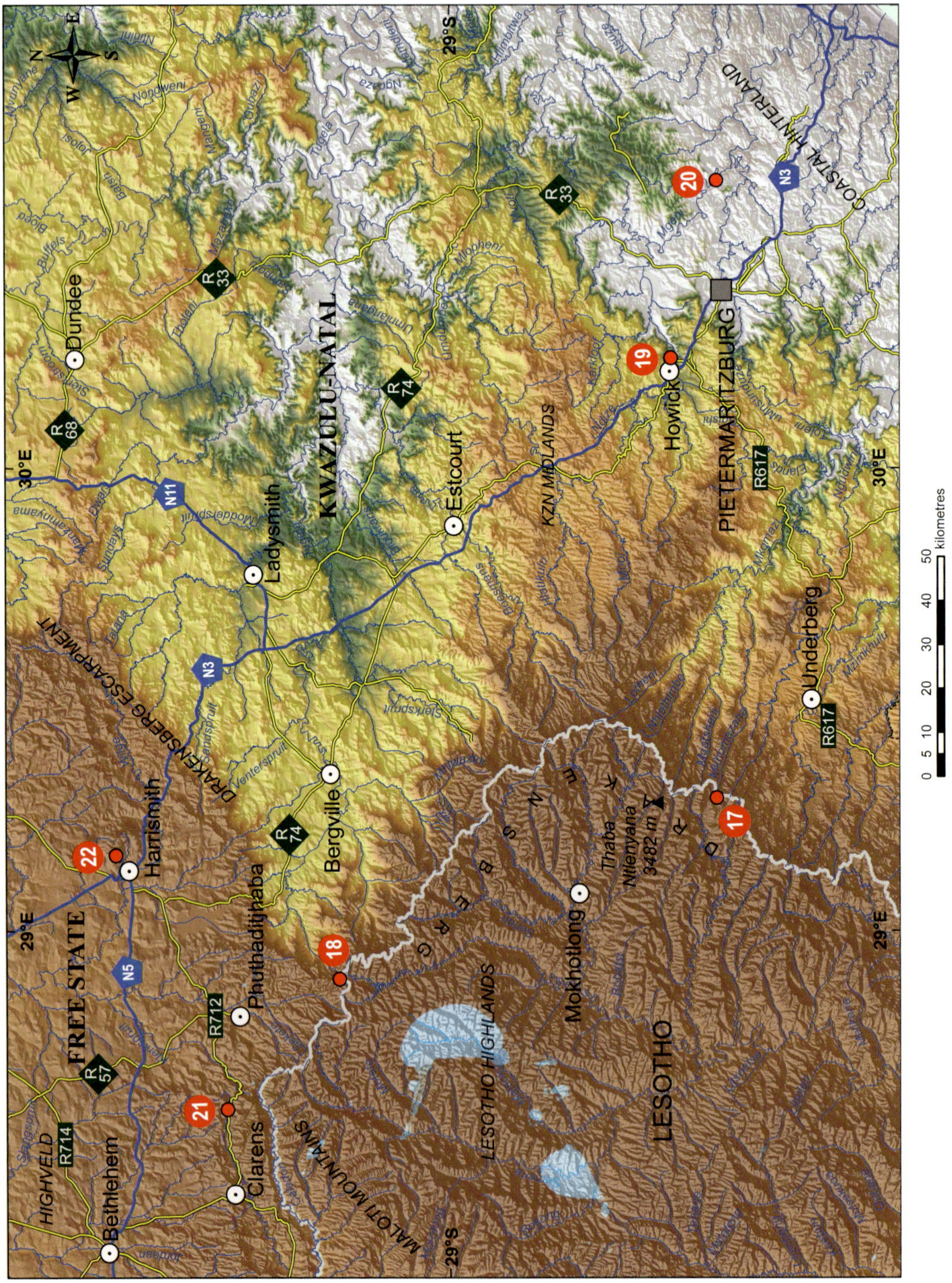

Figure 17: Geological map of central KwaZulu-Natal (KZN) and part of the eastern Free State. Karoo geology makes up most of the region, from the Drakensberg lavas, coloured purple in the west, the Stormberg Group in orange, and Beaufort and Ecca Groups in green and cream respectively. Older granite and gneiss occur in the east, in pink. Note the relative abundance of intrusive Karoo dolerite sills.

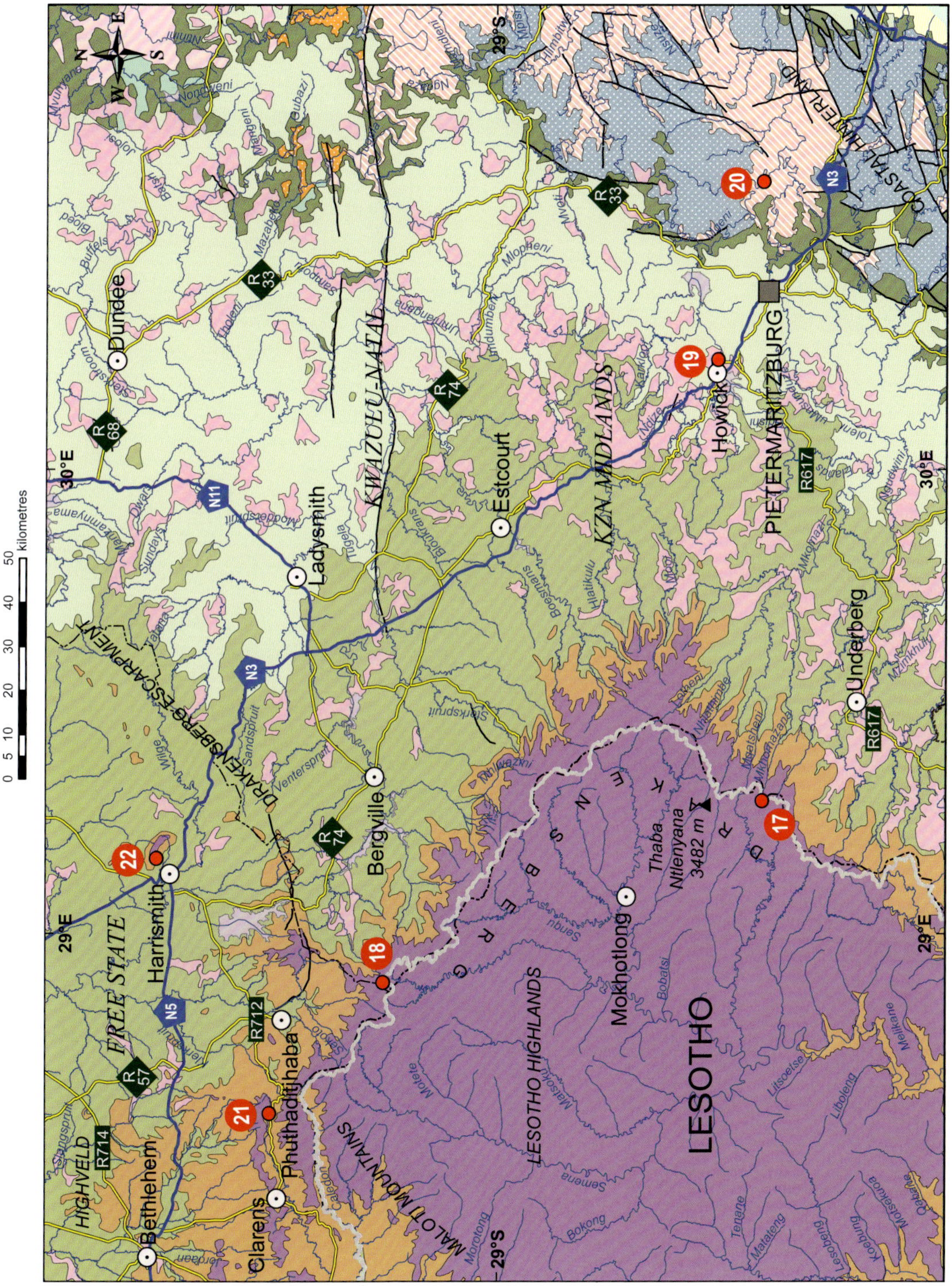

• A stratigraphic key to this simplified regional geological map appears on the inside covers

16 Oribi Gorge

A NATURE RESERVE IN AN INCISED VALLEY

A spectacular and meandering coastal river that has incised deeply into the plateau

The gorge is most impressive from the Oribi Gorge Hotel, where there are numerous viewsites and activities. Here you see the Umzimkulwana River making a deeply incised horseshoe-like meander as it cuts down into the granitic rocks of the Oribi Gorge Pluton.

This outstanding gorge is nearly 30km long and lies along the lower reaches of the Umzimkulwana River, a tributary of the Umzimkulu River, in southern KwaZulu-Natal (KZN). Most of the gorge is within the Oribi Gorge Nature Reserve, a small Ezemvelo KZN Wildlife conservation area marked by almost inaccessible terrain and dense coastal forest, and largely untouched by man.

Location and getting there

- Oribi Gorge is a little off the beaten track, although all the roads are tarred. It starts where the Umzimkulwana and Umzimkulu Rivers meet, about 13km inland from Port Shepstone, and the only access is via the N2 to Kokstad. From Port Shepstone, either turn off after about 10km in the direction of the well-known Oribi Gorge Hotel and continue past the Marble Delta cement factory; or continue a further 18km to Plains and take the turn-off to the Oribi Gorge Nature Reserve.
- GPS (bridge crossing in reserve): 30°42′26″S; 30°16′14″E

Setting the scene

The first record of the gorge is from a hand-drawn map dated 1867, published in John Sheddon Dobie's *South African Journal*, and it was only in the 1890s that the general area became better known. Farming on the Oribi Flats started in the 1900s and in 1928 the Umzimkulwana State Forest, the core of today's 1,917ha nature reserve, was proclaimed. Now almost entirely surrounded by privately owned farmland, the deeply incised gorge presents a surprising spectacle to visitors. It ranges from some 500m–1km wide, and drops 200–300m below the Oribi Flats.

North of Oribi Gorge is the larger Umzimkulu River, which shows the same deep incision through Msikaba sandstone. When the earlier coastal plain was raised, down-cutting by meandering rivers took place, creating the gorges.

The view of Oribi Gorge from Baboon View Walk at the Ezemvelo-KZN Wildlife camp on the southern rim of the gorge. The Oribi Flats are in the distance.

Regional geology

The lower part of the gorge is made of granitic rocks of the **Oribi Gorge Pluton**, a large igneous body (batholith) of the 1,000-million-year-old **Natal Metamorphic Province**. The upper sedimentary part is made of the approximately 400-million-year-old **Msikaba Formation**, which is correlated with the **Witteberg Group** of the **Cape Supergroup**. Previously it was considered a part of the **Natal Group** seen further north (see Valley of a Thousand Hills, pp.138–143) but, on the basis of recent fossil evidence, it is now classed as a separate formation.

Landscape and landforms

The most spectacular feature is the deep gorge that has developed in the Umzimkulwana River Valley. Flanked by the plateau of the intensively farmed Oribi Flats tableland, it provides a superb example of an incised, meandering coastal river valley. Down-cutting was hindered by the layer of hard, horizontal sandstone that is up to 200m thick, but ultimately uplift and strong erosion breached the resistant barrier and cut down through more easily weathered granitic rocks – the result is the spectacular cliffs and deeply incised river valley we see today. The lower slopes are covered by a thick accumulation of blocky sandstone debris. Seen from the hotel, the drop to the river at the large 'horse-shoe' meander is about 300m and the river is only 100m above sea level.

Rock formations

- The Oribi Gorge Pluton is made of coarse-grained granitic rocks of unusual mineral composition, and has attracted considerable scientific attention. It is best seen on scoured outcrops at river level and consists of very large pinkish feldspar crystals set in a finer mass of glassy quartz and dark greenish minerals.
- The Msikaba Formation sandstone is brownish grey when fresh but pale brownish to whitish when weathered. It is cross-bedded (small-scale bedding inclined to the general layering) and gritty in places, and was deposited as clean quartz sand under very stable, shallow marine conditions.

Faulting

A Google Earth image of the area reveals that there are many fault structures that partly dissect the Oribi Flats and have influenced its development. The most prominent fault is in the nature reserve where the east-flowing Umzimkulwana River deviates north for nearly 3km before resuming its eastward course. This is due to a north-south-trending fault that provided a line of structural weakness exploited by the river, and also coincides with the pass through the gorge.

Top: Fresh granite of the Oribi Gorge Pluton is best exposed below the bridge across the Umzimkulwana River Valley in the Oribi Gorge Nature Reserve. Inset: A close-up view of the same coarse-grained granite packed with large, pinkish feldspar crystals.

Msikaba Formation sandstone once formed a thick, horizontal blanket across the area. It is now dissected and is best exposed on vertical cliff faces. Inset: Detail of the well-bedded (layered) nature of the sandstone, as seen along the Oribi Gorge Pass.

Incised valleys and meanders

Deep river gorges are not an uncommon landscape feature along South Africa's eastern coastal margin, where they have formed by a process of uplift causing rejuvenation of the land surface and cutting down of the land surface by rivers. Following the start of the break-up of Gondwana around 150 million years ago, a wetter climate ensued and numerous rivers formed along the eastern side of the subcontinent. Major erosion took place and material was deposited into the young Indian Ocean. This long-lived African cycle of erosion eventually reduced the land surface to a flood plain over which mature rivers flowed to the sea in large loops and meanders. Much later, during the Tertiary Period, renewed uplift of the crust by hundreds of metres changed the base level of erosion and led to the previously sluggish rivers again cutting down and laterally eroding their banks. The earlier meandering river courses became entrenched and existing loops and meanders continued cutting downwards, aiming for the new base level. This pattern of down-cutting is termed an 'incised meander'.

Things to see and do

- **Oribi Gorge viewsites:** The best views are from the area controlled by the Oribi Gorge Hotel. From a flat sandstone plateau you look directly over the precipice to a major loop of the river some 300m below. Named landmarks include The Leaning Tower, Baboons Castle and the Overhanging Rock.
- **Oribi Gorge Nature Reserve Pass:** This scenic 9km-long road takes you from one side of the gorge to the other and provides an ideal stopping place at the river. It is said to follow an old elephant trail which, in turn, reflects an underlying north-south fault line. Fresh, coarse-grained granite can be seen at the road bridge.
- **Nature reserve trails:** There are several popular hikes within the area, leading to scenic places. They all pass through indigenous forest and, in places, reveal the unconformity between hard sandstone and deeply weathered granite.
- **Lake Eland Game Reserve:** This is an extensive private recreational facility located further west at the start of the gorge. Many outdoor activities are offered within and around the gorge, including visiting viewsites, going on hiking trails, game and 4x4 drives, exploring caves and the start of the gorge at Hell's Gate.

Weathered sandstone outcrop seen along Baboon View Walk. The peculiar structure results from differential removal of weathered debris and is known as 'tafoni'.

From the Umzimkulwana River bridge, looking downstream: Oribi Gorge granite is exposed in the river bed and Msikaba sandstone can be seen in the distance.

Baboons Castle, an isolated pinnacle of Msikaba Formation sandstone formed by strong erosion, present near the Oribi Gorge Hotel and Lehr Falls. The Umzimkulwana River can just be seen in the distance.

17

Sani Pass

WORLD HERITAGE SITE

This amazing road climbs through the full sequence of upper Karoo sedimentary formations and Drakensberg lavas to the top of the rugged Lesotho Highland plateau

Looking westwards up the rugged Mkhomazana River Valley in the southern Drakensberg. Sani Pass follows this valley to the narrow neck in the distance, climbing up a precipitous wall of Drakensberg lava. Exposures of typical Clarens Formation sandstone stand out middle right.

Also known as amaNzi'mnyama Pass, this legendary route is South Africa's most hair-raising mountain road – the highest, steepest and the most dangerous, rising almost miraculously more than 1,500m over a distance of about 8km to reach the Lesotho Highlands.

Location and getting there

- Sani Pass is in the southern part of uKhahlamba-Drakensberg Park World Heritage Site. With an incredible average gradient of 1:6, the pass forms part of the gravel road that links Himeville in KwaZulu-Natal and Mokhotlong in Lesotho, and is for the use of 4x4 vehicles only. The pass starts some 25km northwest of Himeville at the ruins of the Good Hope Trading Store, then passes through the border control post, where a passport is required. Re-engineering of the road is in progress and an all-weather surface is planned.

- GPS (border line): 29°35′07″S; 27°17′14″E

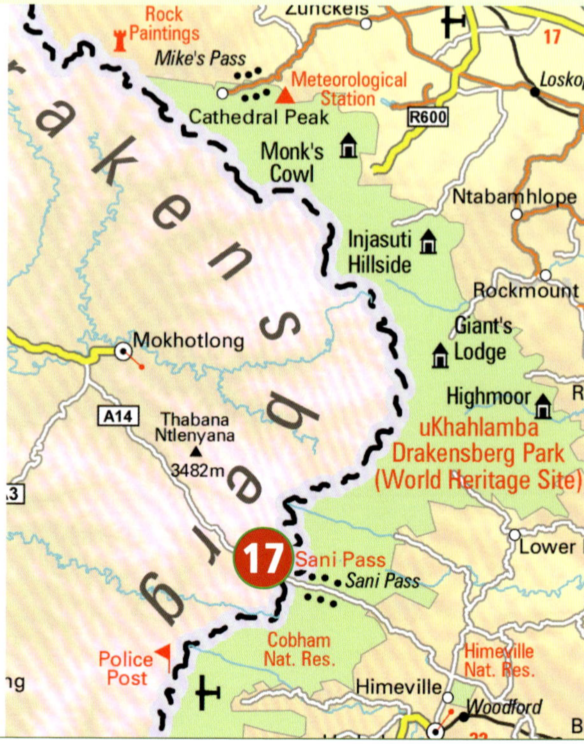

Setting the scene

This major feat of road building ascends the steep wall of the Drakensberg in little more than 8km, from the valley at 1,350m to the Roof of Africa at 2,873m. The pass would originally have been used by indigenous San people, and it became a donkey track leading down the escarpment. In the 1930s it was upgraded to a 2.5m-wide bridle trail

All that remains of the once famous Good Hope Trading Store at the bottom of the pass, a refuge for traders from Lesotho.

The pass winds its way precariously up the mountainside towards the narrow neck in the distance, which coincides with a northwest-trending dolerite dyke, adjacent to the prominent 'knob' seen on the skyline.

and became a lifeline to the people of northeastern Lesotho. The first motor vehicle to negotiate the pass was a Willys Jeep, driven by ex-RAF pilot Godfrey Edmonds in 1948. The hairpin bends were so sharp, and the track so steep and rugged that it took nearly 6 hours of hair-raising effort to complete the journey. Today it can be done in about 30 minutes in a 'soft' 4x4.

Regional geology

The pass is dominated by basalt lavas of the **Drakensberg Group**, the topmost volcanic layer of the **Karoo Supergroup**. These rocks are part of the far more extensive **Karoo Igneous Province** that covered much of southern Africa. Underneath the lava flows is the sedimentary **Stormberg Group**, at the top of which is the **Clarens Formation**.

Landscape and landforms

The pass lies at the head of the strongly incised Mkhomazana River Valley, which has cut back deeply into the Drakensberg Escarpment. The valley is a fine demonstration of aggressive erosion along the escarpment, progressively cutting back ever since the break-up of Gondwana. Erosion is enhanced here by a dolerite dyke that runs northwestwards up the valley and forms the neck at the top of the pass. Sani Top, at 2,873m, is on the edge of the Lesotho Highlands, a relict of an ancient land surface that predates the major break-up of Gondwana from around 150 million years ago.

Figure 18: Schematic cross-section through the stratigraphy of the upper Karoo Supergroup in the Sani Pass and Amphitheatre areas.

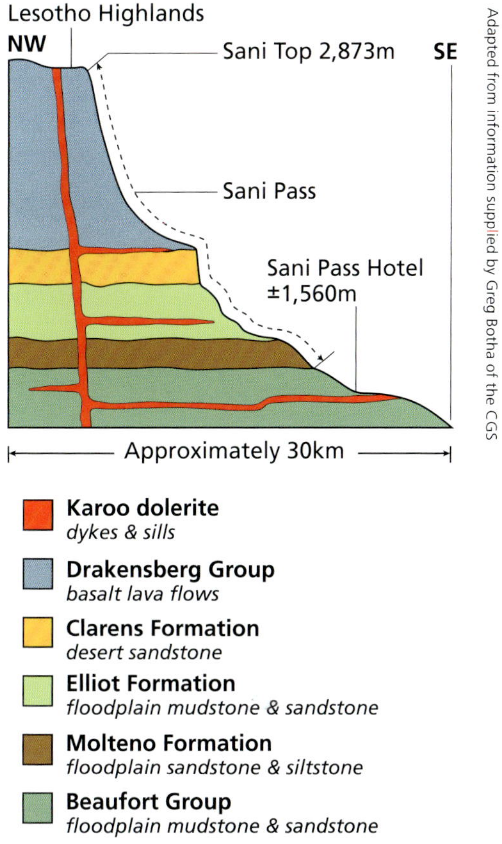

The road near the top of the pass has been cut into Drakensberg lavas; it is in the process of being widened and will in all probability soon become a tarred road.

More about the Drakensberg lavas

Around 180 million years ago a deep mantle plume developed under the northeastern part of southern Africa. Basalt magma was generated by partial melting within the mantle and, where the crust became thinner and began to break along lines of fracturing, it was squeezed up fissures to form lava that poured out across the landscape. During a relatively short period of geological time a vast amount of very liquid basalt lava was erupted at a temperature of around 1,000°C. Sheet-like floods from deep-tapping fissures many kilometres in length eventually covered much of the subcontinent, or were intruded under pressure into existing sedimentary rocks to form a 3D network of Karoo dolerite dykes and sills. The magma formed in southern Gondwana is known as the **Karoo Igneous Province**. This event coincided with the start of the break-up of the Gondwana supercontinent. Today there is not much left of the once far-reaching lava that extended from the Drakensberg to the Zambezi in the north, and to Namibia in the west, and there is only the Lesotho remnant to recall the once piled-up accumulation of lava layers that reached up to 1,500m in thickness.

Rock formations

- Drakensberg lava: Here, this now approximately 800m-thick basalt sequence consists of countless stacked lava flows, varying from 30cm–20m in thickness. The lava flows in the upper part are remarkably similar, constituting the so-called Lesotho Formation, but there is a more variable lava zone at the base. All the basalt lava flows are typically amygdaloidal (contain small cavities filled with mineral deposits).
- Karoo dolerite: A northwest-trending dyke provided the structural locus for the erosion of the valley. It is coarser grained, is not amygdaloidal but is chemically identical in composition to the Lesotho basalt, confirming that it came from the same magma chamber, and could have formed a magma conduit for the overlying lavas.
- Clarens sandstone: This formation lies beneath the Drakensberg lava and forms a prominent scarp along the sides of the Mkhomazana Valley, similar in appearance

Right: An exposure of the blocky Karoo dolerite dyke that cuts through the lavas and effectively created the valley and the neck at the top of the pass. Inset: A close-up view of the dolerite, which has a different, more crystalline texture to the basalt lava, and shows no amygdales.

Looking southeast down the pass and the Mkhomazana River Valley. The road is currently being upgraded.

Weathered basalt shows typical concentric 'onion-skin' weathering.

At least three thin individual basalt lava flows can be distinguished in this exposure, recognised mainly by studying the nature and distribution of the amygdales.

to the Little Berg of the northern region of the uKhahlamba-Drakensberg Park. It is fine grained, creamish to pale greenish in colour and has the same windblown origin (see Golden Gate, pp.144–149).

- Other Karoo Formations: Well-bedded sedimentary rocks, mainly alternating sandstones, siltstones and mudstones of the informal Stormberg Group and upper Beaufort Group, underlie the Clarens sandstone in the incised Mkhomazana River.

Things to see and do

- **Drive the pass:** Without doubt, this is one of South Africa's most exhilarating road trips, and a must for anyone who 'collects' passes. There are several places to stop to admire the view, and also examine the local geology. You can also hike the pass.
- **Attractions on Sani Top:** A number of chalets and a surprisingly good pub and diner exist at the top, with magnificent views down the valley and across KwaZulu-Natal.
- **The Sani plateau and beyond:** On top it is remarkably flat in places. The road, presently being surfaced on the Lesotho side, continues westwards over the 3,350m-high Black Mountain Pass and goes on to Mokhotlong. Nearby is Thaba Ntlenyana – at 3,482m, the highest point in Africa south of Kilimanjaro.

Sign at 2,865m, near the top of the pass and marking the border between South Africa and Lesotho.

18 Amphitheatre

WORLD HERITAGE SITE

Known as uKhahlamba or Barrier of Spears,
this is part of one of the greatest escarpments on Earth

The boulder-filled upper Tugela River, leading away from the 6km-long curved wall of the Amphitheatre that dominates this part of the northern Drakensberg Escarpment.

The Drakensberg Escarpment is a rugged mountain range more than 300km long that separates the rolling hills of KwaZulu-Natal from the highlands of Lesotho, where it is called the Maloti Mountains. The spectacular natural edifice of the Amphitheatre dominates the landscape of the northern part of this almost impassable barrier.

Location and getting there

- The Amphitheatre lies within Royal Natal National Park, which is managed by Ezemvelo KZN Wildlife. To get there take the R74 between Bergville and Harrismith, and turn off near the foot of Oliviershoek Pass. From here the Visitors Centre is about 19km along a tarred road that follows the Tugela River.

- GPS (escarpment top, at Tugela Falls): 28°45′08″S; 28°53′39″E

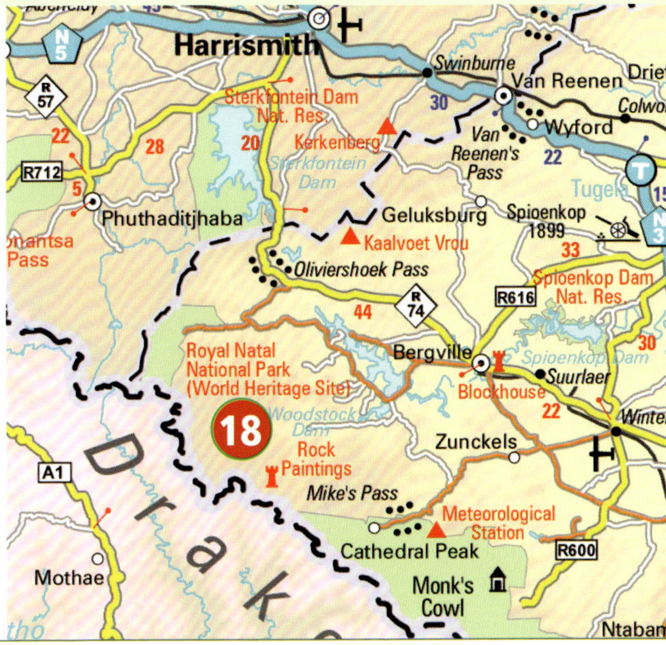

Setting the scene

The KwaZulu-Natal Drakensberg is the highest and longest unbroken mountain range in South Africa, reaching heights of over 3,000m. This is a region of magnificent scenery, considerable natural biodiversity and abundant historic San rock art, located in many cave-like overhangs; in the year 2000, part of the region was inscribed as a Cultural and Natural World Heritage Site, covering an enormous 2,430km^2. The Amphitheatre is a precipitous 6km-long curved and serrated rock wall that rises nearly 1,500m over the Tugela Valley below. Visits to this area will reveal a range of memorable geological and landscape phenomena, to be appreciated over many return trips.

Beacon Buttress (left, 3,124m) and the Sentinel (right, 3,165m) stand out magnificently on the western side of the Amphitheatre.

Regional geology

The uppermost formations of the **Karoo Supergroup** are seen here at their best. On top is a pile of basalt lavas, up to 1,500m thick, called the **Drakensberg Group**, and below that the horizontal and varied sedimentary sequences of the informal **Stormberg Group**. Geological ages range from around 183 million years for the lavas to more than 200 million years for the sedimentary layers (see **Figure 18**).

Landscape and landforms

The Amphitheatre is one of the great escarpments of the world: a rock wall flanked symmetrically by two imposing peaks, the Sentinel (3,165m) and the Eastern Buttress (3,047m). It is a marvellous exhibition of mountain retreat – the rapid 'backward' weathering and erosion of a mountain face, typical of the KwaZulu-Natal Escarpment. There are two landscape domains: the High Berg, which is made of more easily weathered lavas, and the Little Berg, which is topped mostly by massive ramparts of colourful sandstone, the same as that seen at nearby Golden Gate Park (see pp.144–149), although often greyish due to lichen cover. Behind the escarpment edge there is a high plateau, an elevated relict of an ancient weathering surface, where a minor peak called Mont-aux-Sources (3,282m) is the source of three rivers, the Tugela, Elands and Western Khubedu.

Rock formations

- Drakensberg Group: At the Amphitheatre is an immense stack of basalt flows, around 1,000m thick, known here as the Lesotho Formation. These are layers of continental flood basalt that poured out over the late Karoo landscape around 183 million years ago. The thick pile of basalt was built up by countless thin horizontal lava flows that erupted over a relatively short period of geological time lasting about 2 million years, from a network of fissures now filled by dykes. Individual basalt flows, dark greyish or sometimes brownish, range from tens of centimetres to several metres thick, and typically contain countless small, whitish blobs and patches called 'amygdales' (see text box on p.132).
- Clarens Formation: This makes up the prominent sandstone bluffs below the lavas and does much to enhance the attractiveness of the area. The characteristically cream to yellowish,

Yellowish- to grey-weathering Clarens Formation sandstone tops the prominent hills known as the Little Berg. The impressive sandstone-topped hill on the left is known as Dooley (2,056m). Below the sandstone layer are less competent mudstones and thin sandstone and siltstone layers of the Elliot Formation, forming the less steep slopes.

In places the Clarens Formation sandstone has weathered and eroded to form unusual shapes; this one, known as Policeman's Helmet (1,927m), is seen on the way to the Tugela Gorge at the foot of the Amphitheatre.

On the eastern side of the Amphitheatre the Inner Tower (3,044m) forms a formidable peak close to the Eastern Buttress (not shown). In between is an extraordinary needle-like spire known as Devil's Tooth.

The upper Tugela River after it emerges from the confines of Clarens sandstone in the Tugela Gorge.

A broken piece of fresh, fine-grained, whitish Clarens sandstone before it has been discoloured by weathering.

lichen-covered, fine-grained sandstone layer is typically more than 100m thick and generally massive in appearance. It represents windblown desert sand that accumulated immediately before the eruption of the overlying lavas.

- Other Karoo Formations: Below the Clarens sandstone there is the usual conformable sedimentary sequence. It starts with the Elliot Formation, comprising mainly greenish to maroon mudstones, and minor interbedded siltstones and sandstones, followed by the underlying Molteno Formation and upper part of the Beaufort Group, and extensive dolerite sills that form the reddish soil.

What are amygdales?

Amygdales are almond-shaped inclusions in igneous rock or lava. The name comes from the Latin word *amygdala* meaning 'almond', referring to their shape. They are filled with secondary minerals such as agate or chalcedony, and calcite, and sometimes covered with bright green celadonite. They form when very liquid basalt magma is erupted, and the gas and water vapour within the lava is released to produce small, rounded, irregular or tubular gas or steam bubbles. These form mainly along the upper and lower surfaces of the molten lava and become preserved as cavities or vesicles in the solidified rock. Rounded-to-irregular amygdales usually form near the upper surface of a flow; and tube-like pipe amygdales form at the hotter, more liquid base. Over time, mineral-bearing solutions fill the cavities and form secondary minerals (also see Sani Pass, pp.123–127).

Examples of amygdaloidal Drakensberg lava, characterised by small, mineral-filled cavities that are particularly noticeable in the countless river boulders. Amygdales formed as steam or gas cavities within the very liquid lava flows and are usually rounded to irregular but may be pipe-like and branching, indicating upward movement of vapour in the very liquid lava. The bright green coloration occasionally seen is a soft secondary mineral named celadonite.

Things to see and do

- **Drive to the top:** The easy way up is via Witsieshoek Pass and the experience is awesome. Get to Phuthaditjhaba off the R712 from Clarens or Harrismith, and a tarred road takes you to The Sentinel car park at 2,620m. In good weather the final hike to the top can be relatively easy but you need to negotiate vertical chain ladders to reach the escarpment edge at an altitude of nearly 3,000m.
- **Hike to Tugela Gorge:** The 14km route follows the course of the Tugela Valley upstream from Thendele Rest Camp. The gorge is developed within more resistant, massive Clarens sandstone and is littered with enormous boulders. A look upwards gives superb views of different parts of the Amphitheatre, towering over 1,000m above you. When it has rained on top you will have a good view of the plunge of the 950m-high Tugela Falls.
- **Other Royal Natal National Park hikes:** There are many hikes and walks within the park and the surrounding area, all of which provide lessons in geology and the land-forming process in both lavas and sedimentary rocks, including caves (some with San rock art) and waterfalls. The Tugela River has abundant water-worn basalt boulders.

Rod MacLeod

The upper section of Tugela Falls as it drops off the level top of the Amphitheatre and cascades into the Tugela Gorge 950m below. When planning hikes in the area, be aware of the changeable and extreme weather conditions that can occur.

The towering Eastern Buttress (3,047m) in the distance and part of the Little Berg in the foreground, topped by eroded Clarens sandstone, on the path to the Tugela Gorge. You can see why it was called 'Cave Sandstone'.

19

Howick Falls

PROVINCIAL HERITAGE SITE

A spectacular waterfall formed by a dolerite sill at the start of the scenic Umgeni Gorge

The splendour of 95m-high Howick Falls has always been a tourist drawcard for the town, which is located close by, hidden by the trees behind the falls. The view of the falls from town is impressive but the best view (as seen here) is from the Umgeni Valley Nature Reserve.

When driving the N3 past Howick the temptation is to carry on to your destination. However, a short detour to see South Africa's most photographed waterfall is well worth the effort, as is spending extra time to see the equally impressive Karkloof Falls nearby – both falls the result of strong headward erosion by young rivers that have been arrested by resistant rock formations. Howick Falls became a National Monument in 1951 (now a Provincial Heritage Site) to protect this spectacle from development. The Zulu name for the falls is KwaNogqaza or 'place of the tall one', alluding to its considerable height.

Location and getting there

- Howick is located along the R103, just off the N3; the falls are at the edge of the town, and form part of the popular Midlands Meander. Adjacent to the town is the Umgeni Valley Nature Reserve, a small conservation area that covers the valley for 7km downstream, and is managed by WESSA. Karkloof Falls lies 15km by road north of Howick, and about 5km off the road to Rietvlei.
- GPS (town viewsite): 29°29′07″S; 30°14′17″E

Setting the scene

The old wagon road from Durban into the interior forded the Umgeni River just above the falls at Umgeni Waterfall Drift. It was often treacherous after rain, and caused a number of casualties. A village was founded here in 1850; an inn was established and a ferry operated, and the village grew into the town of Howick. Substantial bridges were subsequently built over the strong-flowing Umgeni, which was also tamed by construction of the nearby Midmar Dam. In the 1970s the Umgeni Valley Nature Reserve, now overseen by the Wildlife and Environment Society of South Africa (WESSA), was established in the deeply incised and unspoilt valley below the falls.

Below the falls in the Umgeni Valley Nature Reserve, seen here looking up the gorge towards the town of Howick.

Regional geology

The geology around Howick is dominated by intrusions of **Karoo dolerite** that form the higher ground. They were intruded into flat-lying sedimentary rocks about 183 million years ago, at the same time as the Drakensberg lavas were being erupted. The sedimentary rock formations belong to the **Ecca Group**, the lower part of the **Karoo Supergroup**.

Landscape and landforms

The spectacular scenery around Howick is the result of a dissected and immature landscape, caused by the interplay between harder dolerite sills and dykes and softer Karoo sediments – the latter baked hard (or indurated) in places by the heat from the dolerite, which was intruded as molten magma into the sedimentary rocks at a temperature of around 1,000°C. The process of weathering and erosion has been accelerated by the high rainfall in the region. At Howick Falls the Umgeni River plunges 95m over a dolerite cliff into a gorge, creating the spectacular waterfall – formed where strong headward erosion by the river was arrested by the more resistant thick dolerite sill. This is known as a 'knickpoint' – the position where the river has had its erosion profile re-adjusted.

Rock formations

- Karoo dolerite: The underlying rock formation is a 45m-thick sill of dolerite, overlying hard baked shale. The dolerite sill was formed by molten magma that forced its way into pre-existing horizontal beds of shale and some sandstone layers. At the falls the dolerite sill shows a marked vertical pattern of columnar jointing, a feature that formed after solidification of the dolerite, resulting from contraction of the cooling rock mass (see pp.150–153).
- Karoo dyke: Visible on the far side is a narrow vertical dolerite dyke that was possibly a feeder channel for the Drakensberg lavas, but postdated the sill.
- Ecca shale: About halfway down the cliff there is a 25m-thick, horizontal layer of dark shale (best seen from below) that has been strongly baked and hardened by heat from the dolerite intrusion.

Howick Falls from the town viewsite. Note the vertical structure of the upper sill, a feature called 'columnar jointing' (caused by shrinking of the slowly cooling magma), and the thin cross-cutting dolerite dyke that has been outlined.

Karkloof – another picturesque waterfall

About 15km north of Howick on the Karkloof River is a spectacular waterfall (29°24′01″S; 30°16′44″E) of the same name. The geological situation at Karkloof Falls is reversed, as here the river plunges over horizontal beds of hard sandstone of the **Vryheid Formation**, rather than a dolerite sill. The main drop is 88m over a ledge of resistant, flat-lying sandstone, its hardness attributable to heating by a once overlying dolerite sill, of which only remnants of weathered dolerite now remain. This waterfall formed under similar conditions of aggressive headward erosion and it also represents a knickpoint in the dissected landscape. There is an excellent viewsite about 1km before you reach the falls; from here you will notice the horizontal bedding, but you will need to get closer to get a clearer view of the geology – although getting too close could be dangerous. Further upstream is another cascade, Woodhouse Falls, where the river drops about 25m over the same hard sandstone rocks, although more gradually, over a longer length of river.

Things to see and do

- **View from the town:** Here you have a very good view of the falls and of the geology, and can clearly see the flat dolerite sill and cross-cutting dolerite dyke. Further upstream the same sill forms the rather flat bed of the Umgeni River.
- **Hike down into the gorge:** From the town's viewsite a path takes you to the bottom of the falls where you can see the dolerite sill from a different angle, including spotting another dolerite sill close to river level.
- **Visit the Umgeni Valley Nature Reserve:** You can drive along the upper plateau to the viewsite (about 6km) and also hike along several wilderness trails, including the Shelter Falls Trail in the gorge of the Nkobongo side-stream. There is a magnificent view of Howick Falls from near the camp site. Shelter Falls and Cascade Falls are also within the Umgeni Valley Nature Reserve, on the Nkobongo stream.
- **Visit Karkloof Falls:** Take the Rietvlei road north out of town and drive about 10km to the turn-off. The falls are located on private forestry land some 5km off the main road via a winding drive through timber plantations.

Sign at the town viewsite showing direction and distance of some of the world's more significant waterfalls.

Shaen Adey / Images of Africa

*Howick Falls from the pool below. Note the approximate positions of (**A**) the upper dolerite sill showing columnar jointing, (**B**) the hard-baked Ecca shales showing bedding planes and (**C**) the lower dolerite sill also showing columnar jointing.*

20 Valley of a Thousand Hills

AN ICONIC LANDSCAPE OF KWAZULU-NATAL

One of KwaZulu-Natal's oldest geological terrains and a gigantic example of deep erosion by rivers

Southeast part of the 'Valley' from the top of the road to Nagle Dam. The bare, rounded hills in the distance are granitic rocks of the Mgeni Pluton, which has eroded to form exfoliated granitic domes (similar to Paarl Rock, pp.54–58).

This scenic 'Valley' covering some 1,500km^2 is one of KwaZulu-Natal's most interesting geological regions and a wonderful example of erosion by rivers on a vast scale. Once a 'must-see' landscape, the Valley of a Thousand Hills has been somewhat neglected as a tourist attraction since the construction of the N3 freeway. It lies between Pietermaritzburg and Durban, to the north of Old Main Road (R103), from which long-range views of the landscape open up between Cato Ridge and Hillcrest.

Location and getting there

- The most accessible area of the Valley is that between Old Main Road (R103) and Nagle Dam, and includes KwaZulu-Natal's own version of Table Mountain. Nagle Dam, about 20km further along the tarred road from the Cato Ridge turn-off, is best reached by following Eddie Hagan Drive northwards; this takes you off the plateau, and it is signposted and tarred all the way. From Pietermaritzburg an alternative route is via Table Mountain Road, which takes you eastwards to the accessible western side of Table Mountain.
- GPS (Nagle Dam diversion weir): 29°35′09″S; 30°37′27″E

Setting the scene

The Umgeni and its tributary rivers make their convoluted way through the jumbled landscape of the Valley to the sea more than 60km away. At Nagle Dam the Umgeni River is impounded in a U-shaped meander, while below the dam the main tributary is the Msunduzi River, famous for the annual Duzi Canoe Marathon. Between these two rivers lies the remarkably flat-topped massif of KwaZulu-Natal's Table Mountain, while to the south, above the sandstone escarpment, runs Old Main Road, which is part of the famous Comrades Marathon route and passes through familiar-sounding places like Hillcrest, Drummond and Botha's Hill. The road to Nagle Dam winds through what has become a burgeoning low-density residential development, notably at lower elevations.

Looking northwest across the Valley with KZN's Table Mountain in the distance – an isolated remnant of the once extensive covering of horizontal Natal Group sandstone.

Regional geology

There are two very different components to the geology of the Valley of a Thousand Hills.

Firstly, there's an extensive inlier of coarse-grained intrusive granites and deformed older gneisses that are part of the **Natal Metamorphic Province**. Most of the Valley is underlain by the enormous **Mgeni Pluton**, a large complex body of granite that intruded the **Mapumulo Group** gneisses (see also Oribi Gorge, pp.118–122). These formations (1,250–1,000 million years in age) were part of an enormous fold mountain range that developed by subduction (the process at the boundaries of tectonic plates, where one plate moves under another) along the southern margin of the Kaapvaal Craton, and they are closely akin to very similar metamorphic formations found in distant Namaqualand.

Secondly, there's the horizontal sandstone formation of the **Natal Group** that was deposited around 490 million years ago, and which forms the more resistant plateau and isolated remnants such as KwaZulu-Natal's Table Mountain. This formation is correlated in time with the sandstone-rich **Table Mountain Group** of the Cape, and it is partly covered by a thin layer of the **Karoo's Dwyka** glacial sediment, deposited around 300 million years ago, and the lowermost unit in the **Karoo Supergroup**.

Landscape and landforms

The rugged Valley is a fine example of river erosion, currently in a youthful state. Enhanced by higher rainfall and uplift since Cretaceous times, the landscape is dominated by rejuvenated and deeply eroding rivers. These, together with their countless feeder streams and gullies, have formed a complex pattern of interlocking spurs in a terrain that is underlain by deeply weathered and mostly soft granite. Seen in the distance to the east of the road to Nagle Dam are several large bare hills of more resistant granite, one of which is informally known as Old Baldy.

The southern end of Table Mountain. The rather flat top was worn down by Dwyka glacial erosion. Inset: A close-up look at the well-bedded, reddish-weathering Natal sandstone that makes up Table Mountain and the plateau further south.

Erosion is controlled by the Umgeni and Msunduzi Rivers, their convoluted courses indicating that they once meandered across a flat plain; when this plain was subsequently uplifted, the rivers cut down into the raised surface, creating deeply incised meanders.

In the south, horizontal Natal Group sandstone still forms the high ground and can be seen as cliffs close to Old Main Road, while the vertically sided Table Mountain forms a magnificent mesa, which has become geologically detached, through erosion, from the main plateau-forming sandstone. This thick sandstone would have formed a continuous layer before coastal uplift and subsequent dissection and erosion by rejuvenated rivers took place.

Rock formations

- Dwyka Group: Only seen on the top of KwaZulu-Natal's Table Mountain, this is the basal remnant of the once overlying Karoo formations; the rest of the Karoo Supergroup has long since been stripped off by erosion. From a distance it is seen as a grassy capping but you need to get on top of the mountain to see the rocks.
- Natal Group: This is best seen as the road leaves the plateau escarpment near Cato Ridge; it consists of resistant flat-lying, reddish-weathering sandstone, and also forms the impressive layered cliffs that surround Table Mountain.
- Mgeni Pluton: This large, complex intrusion is made of very coarse-grained granite, typically consisting of very large crystals of pinkish feldspar set in a much finer mass of biotite and quartz. Fresh granite is seen in a road cutting at the start of the valley road but lower down in the Valley the granite is deeply weathered and easily eroded.
- Mapumulo Group: These are the oldest rocks in the region and comprise layers of strongly deformed, light-coloured quartz-feldspar gneiss, biotite gneiss and dark amphibolite. They are well banded and have a distinct northeast structural trend, and dip to the northwest. They are best seen close to Nagle Dam.

An outcrop of the Mgeni Pluton along the roadside. Inset: A close-up of the coarse-grained, 'megacrystic' granitic rock comprising glassy quartz, very large pinkish feldspar crystals and abundant fine-grained biotite.

An exposure of banded Mapumulo Group gneiss near Nagle Dam. This group of rocks is the oldest and most deformed in the Natal Metamorphic Province.

A pegmatitic vein that formed in the Mapumulo gneiss, consisting mostly of very pink feldspar.

In places the Mgeni Pluton granite is deeply weathered. Here only residual core-stones are preserved.

A horseshoe bend (an incised meander) in the Msunduzi River close to Nagle Dam indicates that the river once meandered slowly across a flood plain before uplift and later down-cutting by river erosion.

Fresh granite of the Mgeni Pluton is well exposed in places in the incised Umgeni River. Inset: The characteristic 'megacrystic' texture of this coarse-grained granitic outcrop.

KZN's Table Mountain and its glacial heritage

Known in isiZulu as Mkhambathini, meaning 'the place of the acacias', KZN's Table Mountain is the most prominent landmark in the region, forming a spectacular natural sandstone edifice that rises to more than 900m, about 500m above the rivers far below. Narrowing wedge-like to the north, it is oriented northwest, about 3.5km long and up to 1.7km wide in the southeast. Although it appears almost flat-topped from afar, in reality the top is a gently undulating plateau of pristine grassland that reaches an elevation of 959m.

This mesa is surrounded by cliffs of reddish Natal Group sandstone, and the encircling wooded slopes of rock-fall are underlain by granite-gneiss of the Natal Metamorphic Complex. The central ridge of the mountain top is covered by a layer of sedimentary rock of glacial origin, the top of which was scoured by ice; the resulting well-preserved glacial striations indicate that the Dwyka ice sheet moved across the landscape from the northeast. The glacial sedimentary rocks are now deeply weathered to form reddish soils.

On top of Table Mountain you can see diagnostic evidence of glacial striations scoured into Natal Group sandstone by a Dwyka ice sheet. From on top there's a fine view of Nagle Dam, built in a loop of the meandering Umgeni River (not shown here).

Greg Botha

Things to see and do

- **Drive to Nagle Dam:** The 20km-long winding, scenic drive will take you from Cato Ridge down into the Valley of a Thousand Hills, across the Msunduzi River and past some of its meanders on the road to Nagle Dam on the Umgeni River; along the way you will see varied geology and the deeply incised landscape.
- **Visit Nagle Dam:** Run by Msinsi Resorts, this dam and game reserve is less than an hour's drive from Durban or Pietermaritzburg, and provides a range of outdoor recreational activities, including game-viewing, boating, fishing and hiking. You will be able to see the local geology, with the impressive Table Mountain in the background.
- **Hike up Table Mountain:** A walk up the mountain is possible from the western side, provided you have permission from local authorities; it makes for an invigorating climb and provides wonderful views of Nagle Dam and the Valley to the east.

21 Golden Gate

A SOUTH AFRICAN NATIONAL PARK

Spectacular Karoo sandstone scenery and the world's oldest dinosaur eggs

No scene is more evocative of the eastern Free State than that of The Sentinel near to Glen Reenen Rest Camp. A short hike will take you to the top of this Clarens Formation sandstone feature. In the winter months snow is a common occurrence, seen here covering the higher peaks in the distance, which are composed of Drakensberg lavas.

This is one of South Africa's most iconic landscapes, well known for its impressively coloured sandstone scenery, the best of which lies within the Golden Gate Highlands National Park. It is part of the Rooiberg range that forms the foothills of the Maloti Mountains, and is a western extension of the KwaZulu-Natal Drakensberg.

Location and getting there

- The national park lies some 20km east of Clarens on the R712, which follows the course of the west-flowing Little Caledon River that has its source at nearby Wodehouse Peak (2,438m high). This scenic public road runs directly through the park, where it reaches 2,041m at Lichens Pass and continues on to Phuthaditjhaba (previously Witsieshoek) and to Harrismith.
- GPS (Sentinel Buttress top): 28°30'10"S; 28°36'51"E

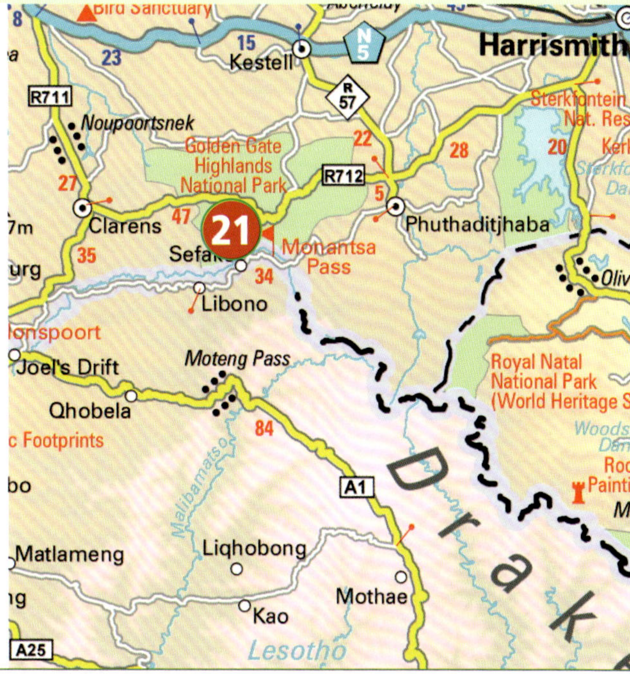

Setting the scene

The name Golden Gate was given in 1875 to the spectacular western sandstone portal, which glows yellowish and reddish in the setting sun. Although the area had long been enjoyed for its outstanding natural beauty, the park only came into existence in 1963, and it took on the earlier evocative name. It is now approximately 120km^2 in area and is managed by SANParks.

Regional geology

The formations represent the uppermost part of the **Karoo Supergroup** and comprise, from top down, the **Drakensberg Group** and the upper part of the sedimentary **Stormberg Group**, which predates the Drakensberg lavas. There are also numerous **Karoo dolerite** intrusives around 183 million years in age. The terrain is predominantly steep, open montane grassland.

Landscape and landforms

Golden Gate is located at the watershed between the Wilge-Vaal and Caledon-Orange river systems. The extreme climate – high rainfall, hot summers and freezing winters – has taken its toll on the landforms, resulting in eroded valleys and dissected highlands. The highest peak is Ribbokkop (2,829m) close to the Lesotho border. Most impressive are the spectacularly coloured vertical sandstone landforms, providing textbook examples of weathering and erosion. Cave-like overhangs are common, caused by differential sandstone weathering and undercutting. While the iconic tourist image is of The Sentinel buttress (Die Brandwag) standing guard over the Glen Reenen Rest Camp, there are many more small-scale geomorphological features and landforms to see, such as calcareous concretions, surface spalling, polygonal cracking and lattice-like weathering.

At Rooidraai an exposure of reddish Elliot Formation mudstone is cross-cut by a Karoo dolerite sill, and shows 'baking' of the adjacent mudstone in the bleached contact zone. The fossilised dinosaur eggs came from very near here.

Rock formations

- Drakensberg Group: Up to 600m of dark grey, amygdaloidal basalt lava flows make up the high ground and peaks, and have produced blackish soil covered by tough mountain grass. This basalt formation is the result of lava flows that repeatedly poured out across the sandy landscape, covering much of southern Africa (see Amphitheatre, pp.128–133). Numerous dykes and sills of Karoo dolerite have also intruded, and a volcanic pipe is reported near Cathedral Cave, intruding through sandstone.
- Clarens Formation sandstone: Below the Drakensberg basalt is what was previously called the 'Cave Sandstone': uniformly fine grained and yellowish white when fresh, and massive, without obvious internal structure. Towards the base there is a zone of pink to reddish sandstone, probably due to oxidation at the time of deposition around 190 million years ago. The Clarens sandstone is up to 150m thick and, in places, large aeolian cross-beds (units that are composed of moderately inclined layers) reflect its windblown origin in a desert of Saharan proportions that existed at the closing stage of Karoo sedimentation.
- Elliot Formation: Below the Clarens sandstone, and seen only in the 'lower' river valleys, this formation consists of flat-lying beds of red mudstone and siltstone, and lesser, fine-grained yellowish sandstone. Previously known as the 'Red Beds', this formation was deposited on an alluvial flood plain around 200 million years ago, and in places contains abundant dinosaur remains, including nesting sites and fossil eggs (see opposite page).

Fossilised dinosaur eggs

Golden Gate is well known for vertebrate fossils, evidence of life in a rich ecosystem before the entire subcontinent became a desert and was then overwhelmed by flood basalts. In 1977 fossilised eggs were discovered at a road cutting near Glen Reenen Rest Camp, close to the Rooidraai viewsite: an ancient nesting site containing very fragile, intact dinosaur eggs entombed in mudstone of the Elliot Formation below the Clarens sandstone cliffs. These eggs are from a large, vegetation-eating sauropod dinosaur known as *Massospondylus* that grew to a length of around 5m. Since then, at least 10 more nests containing 34 eggs have been found from the world's oldest dinosaur communal nesting site. There are also footprints of young dinosaurs that had not yet left the safety of the nesting site, which was covered by muddy sediment during a flash flood around 200 million years ago, preserving the tracks forever. Recent scientific work on the round eggs, slightly larger than those of a domestic hen, has revealed minute, bony, embryonic dinosaurs within. The fossilised eggs are housed at the Bernard Price Institute of Palaeontology at the University of the Witwatersrand. A dinosaur interpretation centre at Golden Gate is planned.

Prof R Reizs, University of Toronto

Prof R Reizs, University of Toronto

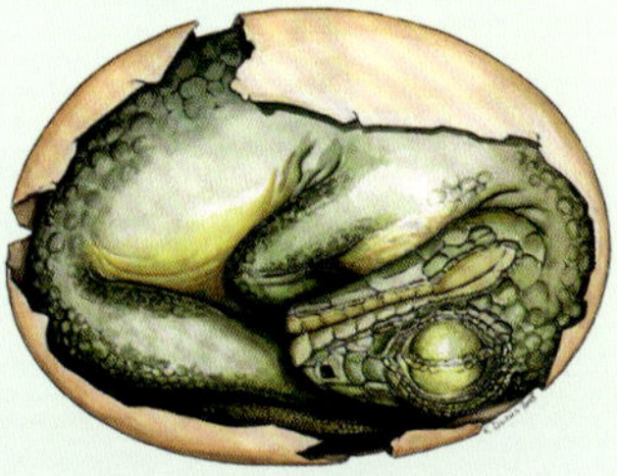

Prof R Reizs, University of Toronto

From fossil diorama at Albany Museum

Above: A primitive leaf-eating dinosaur, Massospondylus, *is being attacked by smaller, flesh-eating dinosaurs, perhaps while protecting a nesting site. In the background, dunes of the overlying Clarens Formation are starting to form. Note the two small, shrew-like* Megazostrodon, *among the earliest-known mammals, one foraging in a bush, the other fleeing (left and bottom right respectively). Top gallery: Two representations of* Massospondylus *embyros: left shows perfectly preserved fossil embryonic bones, and centre is an artist's reconstruction of an egg found at the Rooidraai site. Right inset: An excavation crew recovering rock containing fossil eggs at Rooidraai.*

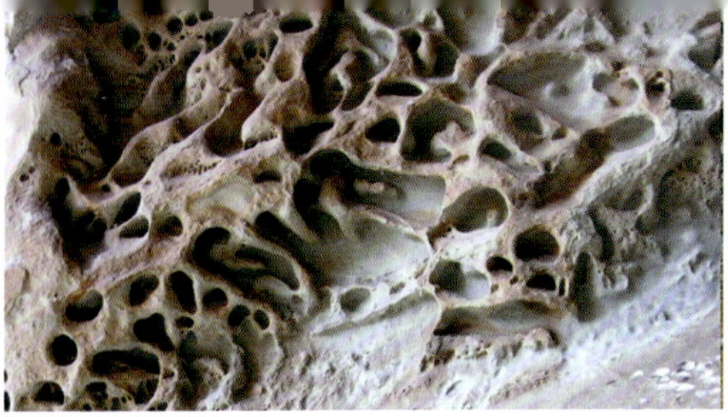

Above inset: In the caves are many examples of 'honeycomb weathering' also known as 'tafoni', a feature explained by differential weathering in granular rocks.

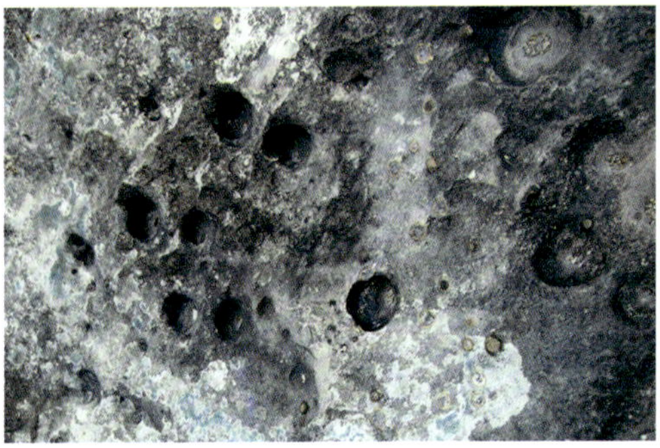

Above: On the short Holkrans Trail behind the Golden Gate Hotel there is a very large and interesting overhang formed in Clarens sandstone, plus several other smaller caves. There are numerous small-scale geological features to be seen, including a pinkish zone probably close to the contact with underlying Elliot red shale.

Left inset: Parts of the Clarens sandstone contains many small rounded nodular-like concretions or their weathered-out holes, formed long after deposition by a mineral-concentrating chemical process. Their origin is not well understood.

At the base of The Sentinel a thick, dipping dolerite sill crosses the main road. Look out for the columnar jointing formed by slow cooling. Inset: At road level there are excellent exposures of slightly porphyritic dolerite, which also shows columnar cooling joints; this is the same dolerite sill that has 'baked' and preserved the sandstone layers that have resisted erosion at The Sentinel.

Things to see and do

- **Climb The Sentinel:** This is a must-do and provides a wonderful, all-round view. The route via Glen Reenen takes you across the thick dolerite sill, the heat of which helped to harden the sandstone. Look for the same dyke on the main road, where it displays columnar jointing.
- **Drive the loops:** The 6.7km Blesbok Loop takes you above the Clarens sandstone into the Drakensberg lavas and provides breathtaking scenery; and the 4.2km Oribi Loop gives a magnificent view of the distant Drakensberg.
- **Hike to Holkrans:** This takes you to a large, cavern-like overhang in Clarens sandstone where there are many interesting small-scale geological structures. There are also short walks along the base of Mushroom Rock and to the cleft of Echo Ravine.
- **Stop at Rooidraai:** In a road cutting about 1km east of Glen Reenen there is a magnificent exposure of friable, reddish Elliot mudstone intruded by two relatively thin dolerite dykes cutting through in different directions. Think about the dinosaurs that once thrived here, on the muddy flood plain.

The 6.7km-long Ribbokkop Loop takes you into the high ground to the south, including Langtoon Dam, from where you can see the highest peak, Ribbokkop (2,829m), in the far distance on the right.

From the 4.2km-long Oribi Loop there are good all-round views, including the distant snow-capped Drakensberg. The highest peak, to the left, is The Sentinel (3,165m), part of the Amphitheatre. Clarens Formation sandstone is well exposed in the middle distance.

22

Platberg

HARRISMITH'S ISLAND MOUNTAIN

A fine example of a large Karoo inselberg with a capping of columnar jointed dolerite

Platberg seen from its western side, standing proud above the grassy plains of the Free State. From this vantage point the overlying columnar-jointed Karoo dolerite sill and partly covered Clarens Formation sandstone can be seen, and the surrounding scree-covered slopes.

The impressive flat-topped mountain or 'mesa' near Harrismith, known as Platberg, is usually the signal for a welcome road stop when you are driving on the N3. How many times have you driven past this enormous inselberg and wondered what it's really about?

Location and getting there

- Platberg lies northeast of the N3 but to get there you need to go into Harrismith, then pass under the N3 to the Platberg Nature Reserve. You will have to pay to get into the reserve and then there are several options: take the Mountain Drive along the southern slope; go on an exhilarating 4x4 drive to the top via Donkey Pass; or simply hike to the top.
- GPS (start to Donkey Pass): 28°16′51″S; 29°12′01″E

Setting the scene

The geological map of the area shows that Platberg is capped by an outlier of Drakensberg basalt lava, as are the nearby smaller peaks of Mooihoekkop to the north and Loskop to the northwest. Platberg shows one of the most impressive occurrences of columnar jointing to be seen in South Africa. This not uncommon feature is similar to the thick dolerite sill that makes up the rugged landscape at the Valley of Desolation (see pp.106–109).

Regional geology

Platberg is positioned at the very top of the **Karoo Supergroup**. Here a flat-lying, sill-like body of **Karoo dolerite** overlies whitish sandstone of the **Clarens Formation**, which is some 190 million years old. Encircling the mountain is a thick scree slope formed by the accumulation of weathered and

Remnants of collapsed dolerite columns on Donkey Pass. Note the very distinct vertical joint surfaces that formed during slow contraction of the already solidified magma.

A typical example of distinctly porphyritic dolerite outcrop seen on the top of Platberg. The porphyritic nature of the rocks is shown by the small whitish flecks of early crystallising feldspar.

A close-up of some of the large, angular, collapsed dolerite columns close to Donkey Pass; some are several metres in length. Originally, they stood upright.

Going up Donkey Pass you will see this cave-like overhang formed within whitish Clarens sandstone, overlain by the Karoo dolerite sill, which shows considerable vertical columnar jointing.

Because of their abundant jointing, the dolerite columns are prone to collapse as shown by this fairly recent rock avalanche, coming from the dolerite sill higher up.

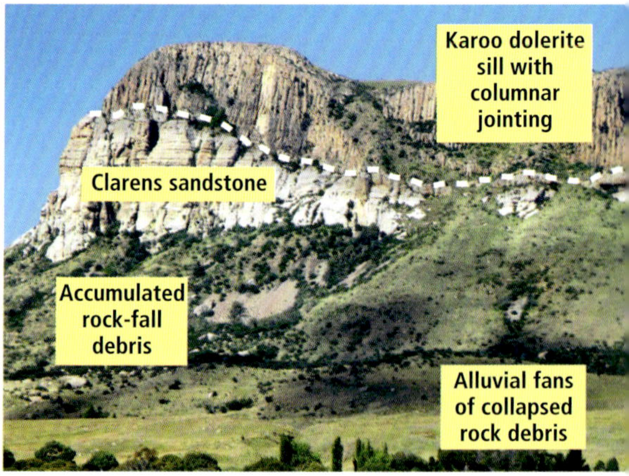

To the northwest of Platberg an isolated hill known as Losberg (called Everest by some) reveals the marked transgressive nature of the sill cutting across beds of horizontal Clarens sandstone (dashed white line). Alluvial fans made of collapsed rock debris are shown on the right.

eroded rock debris over many millions of years. A major dolerite dyke curves past the eastern side of Platberg and Mooihoekkop, and was possibly the feeder channel for the overlying dolerite sill.

Landscape and landforms

Platberg is an inselberg, literally an 'island mountain'. It is geologically linked to the Drakensberg Escarpment (to which it would once have been connected), and is a fine example of a *mesa* or 'table mountain'. The surrounding plain has been completely eroded down and is part of the long-lived African erosion surface that formed following the break-up of Gondwana, leaving the mountain an isolated inselberg, a mere remnant of an earlier land surface. It is aligned more or less northwest, is about 8km long, up to 2km wide and rises more than 500m above the plain. It reaches a height of 2,395m above sea level. The full reason for its preservation is not clear, but is at least partly due to its thick cover of dolerite.

Rock formations

- Karoo dolerite: The sill-like, dark capping is impressively columnar jointed, and is estimated to be up to 40m thick. This is the same rock type that forms countless sills and dykes in the Karoo and is from the same magma that made the Drakensberg lavas around 183 million years ago (see Amphitheatre, pp.128–133).
- Clarens Formation sandstone: Below the dolerite capping, this fine-grained, whitish sandstone is of windblown (aeolian) origin and is best known from the Golden Gate area (see pp.144–149). It dates from the final stage of desert deposition at the end of Karoo sedimentation, before Karoo volcanism took hold. Well known for its overhanging caves, this formation used to be called the 'Cave Sandstone'.

What is columnar jointing?

The world's best-known columnar jointing is at Giant's Causeway in Northern Ireland, a World Heritage Site and major geotourism destination. Here, around 40,000 mostly hexagonal, interlocking black basalt columns up to 12m high are wonderfully preserved. When a thick basalt flow or dolerite sill has cooled and solidified, columnar jointing may develop within the slowly cooling, tabular, homogeneous igneous rock mass. During cooling of such an igneous rock, stresses are set up by systematic and uniform contraction of the cooling mass and a pattern of polygonal cracks develops at right angles to the cooling surfaces. Cooling takes place more or less equally in all directions and, ideally, a hexagonal joint pattern is formed – although perfect symmetry is not always possible and other sorts of polygons also develop. Platberg offers one of the best developed examples of columnar jointing in South Africa but there are many other examples in Karoo dolerite sills and Drakensberg basalt.

Things to see and do

- **Go up Donkey Pass:** This rough concrete strip-road (4x4s only) takes a direct route to the top via a narrow kloof on the southeastern side of the mountain. On top there is an undulating plateau of pristine montane grassland, from where you will have a panoramic view of the surrounding country, and you will be able to see:
  - **Dolerite outcrops:** Mostly subdued but well exposed at ground level, their shape reflecting the underlying polygonal columnar jointing.
  - **Massive dolerite columns:** There are many alongside the Donkey Pass track, some standing, some collapsed, and others as remnants. Close-up, the columns are large, around a metre or more across, and many metres high when standing.
  - **Clarens sandstone:** Typically massive and of whitish colour, and in which a characteristic cave-like overhang has formed, so you can see why it was originally known as 'Cave Sandstone'.
- **Look at Loskop:** This hill (also known as Everest by some) is a remnant that was once connected to Platberg. Here you can see that the capping body of dolerite clearly cuts across the horizontally bedded Clarens sandstone, providing good evidence of an intrusive origin for the dolerite.
- **See a fossil tree:** Lying outside Harrismith's town hall there is a complete and impressive 30m-long fossil tree trunk, well protected by fencing. Presumably collected nearby, it is an ancient Karoo tree, of the genus *Dadoxylon* and is around 250 million years old.

23

Vredefort Dome

WORLD HERITAGE SITE

The world's largest and oldest recognised complex meteorite impact site

This now disused rock quarry at Leeukop, some 4km northwest of Parys, was once mined for dimension stone. It reveals excellent three-dimensional exposures of Parys granite-gneiss cut by thick intrusions of the once enigmatic and complex rock called 'pseudotachylite breccia'. This is one of three small 'satellite sites' of the much larger UNESCO World Heritage Site. Other rock quarries in the area show similar material.

When meteors – small pieces of planetary debris – enter the Earth's atmosphere they mostly burn up by friction and are called 'shooting stars'. Occasionally, however, small pieces may survive to land on the Earth's surface as meteorites. A little over 2,000 million years ago a rare, mountain-sized meteorite (or asteroid), moving at great speed, struck near to where the small town of Vredefort is today, with catastrophic results.

Location and getting there

- The Vredefort Dome, as both the geological and geographical features are known, covers parts of northern Free State and eastern North West provinces. It includes the town of Parys, which is the regional tourist centre about 120km south-southwest of Johannesburg (a 1.5-hour drive). A number of roads get you there, the easiest route being via the N1, turning west to Parys on the R59. There are several alternative main roads leading to Parys or Vredefort.

- GPS (Vredefort Interpretation Centre): 26°59′17″S; 27°22′19″E

Setting the scene

Only in the late 1990s, after years of scientific debate and study, was it finally confirmed that the large, multiple, ring-like geological structure at Vredefort was formed 2,023 million years ago by a catastrophic meteorite impact. Recognised as the oldest and largest complex meteorite impact site on Earth, it is surrounded by a broad crescent of rugged mountains in the scenic northwestern part of the structure close to the Vaal River. It was, in part, declared a World Heritage Site in 2005, and includes three small satellite sites. The massive crater that was produced has now completely eroded away but the impact, which left considerable diagnostic evidence of the event, was of fundamental importance in preserving the Witwatersrand goldfield, the world's largest deposit of gold. It is also considered to have been the world's greatest single known energy-release event.

Visitors to Parys are welcomed by polished and roughly quarried blocks of attractively coloured Parys granite-gneiss – once a popular dimension stone traded as 'African Juparana' and used in various public buildings around Gauteng.

A closer-up view of the enigmatic veins of pseudotachylite breccia (**A**) *at Leeukop Quarry formed at the time of the meteorite impact and resulting from superheated friction or shock. As shown, it consists of many different sizes of rounded granite-gneiss clasts set in a dark glassy matrix of melted rock, within streaky Parys granitic gneiss* (**B**).

Regional geology

The Vredefort Dome shows itself as an approximately 90km-wide, semicircular geological structure located centrally on the Kaapvaal Craton; and airborne geophysical surveys reveal that the ring structure is continuous underneath much younger Karoo cover rocks in the south and east (see **Figure 21**). In the centre, or **Core Zone**, there is a 45km-wide expanse underlain by 3,200-million-year-old **Basement granitic gneiss**, consisting of an outer zone of more colourful Parys granite and a complex inner zone of high metamorphic-grade rocks from the deeper crust. Around this core is an approximately 25km-wide **Collar Zone**, comprising steeply dipping sedimentary and volcanic formations of the pre-impact **Witwatersrand**, **Ventersdorp** and **Transvaal Supergroups**. The Vredefort Dome is located approximately in the centre of the roughly oval-shaped Witwatersrand Basin, which probably owes its preservation to the down-warping of the crust due to the impact.

Landscape and landforms

The landscape around the Vredefort Dome is part of the Highveld plateau, and forms the Vredefort Mountain Land (or Bergland) on the Dome's western and northern sides. Here the crescent-shaped mountains are strongly geologically controlled by upturned Witwatersrand, Ventersdorp and Transvaal strata, through which the westward-flowing Vaal River has carved a scenic course after passing through granite-gneiss around Parys. On the southern and eastern sides of the Dome the much flatter plain is partly covered by significantly younger Karoo strata and intrusive Karoo dolerite sills overlying granite-gneiss and upturned sediments.

An example of the Vredefort granophyre close to the town of Vredefort, thought to be a remnant of the original impact melt preserved as a dyke-like structure. It contains abundant small inclusions of pre-existing rock types.

A smooth outcrop of pseudotachylite breccia at the Salvamento Quarry near Leeukop excellently shows the complex nature of the breccia cutting across a granite-gneiss pavement. The dark material is glassy pseudotachylite.

This is what the Parys granite-gneiss looks like before it is exposed by quarrying, as seen close to the main road east of Parys.

An example of streaky pinkish Parys granite-gneiss from a test quarry near the new Interpretation Centre at Vredefort.

These famous shatter-cone structures in quartzite, seen at Schoemansdrift across the Vaal River west of Parys, are absolutely diagnostic of a meteorite impact origin.

The Amazon quartz pebble conglomerate found north of Venterskroon is part of the Witwatersrand's Kimberley Formation, seen in the upturned Collar Zone sequence. From the 1880s–1930s this 'reef' was the focus of gold prospectors in the small Venterskroon goldfield.

Rock formations

Certain of the widespread rock formations that provide important and diagnostic examples of the Earth-shattering impact event (see text box below and diagrams pp.159, 161) are listed here.

- Vredefort granophyre: A few hard, dark, narrow dykes cutting through granite-gneiss contain abundant small fragments of earlier rocks, such as shale and quartzite. Scientific work, including finding chemical traces of the ancient meteorite, has determined that these rocks represent the original impact melt that penetrated cracks in the floor underneath the original impact crater.
- Parys granite: This is Archaean Basement granite-gneiss of the outer Core Zone, which contains variable proportions of streaky pink to reddish gneissic banding and was used extensively in the polished-stone industry. There are numerous abandoned stone quarries around Parys.
- Pseudotachylite breccia: This enigmatic rock is a feature of the Vredefort Dome and is best seen in exposures of quarried Parys granite. It comprises irregular veins of black glassy material containing rounded inclusions of the granite-gneiss, and was formed by intense friction or 'shock-melting' at the time of impact.
- Witwatersrand Formations: The full sequence is present but thinner, shattered and now steeply dipping or even overturned. These are the same rocks that outcrop along the Witwatersrand, including conglomerates, confirming that they were pushed up by the post-impact rebound.
- Shatter-cones: These rare, small-scale geological features with fine, radiating, cone-like (or fan-like) striations are found worldwide, and are associated exclusively with meteorite impact structures and with underground nuclear explosions. Their formation is due to the impact shock wave moving through certain types of rock.

The Vredefort impact event

Improved understanding of the impact craters on the Moon has enhanced our understanding of this Earthly impact event. The diagram opposite shows the sequence of events that unfolded when a meteor approximately 10km in diameter slammed into planet Earth. It was travelling at over 10km per second (over 36,000km per hour), and its enormous kinetic energy was released on impact when it exploded deep in underlying Basement granite-gneiss. Hundreds of millions of tons of rock and dust were blasted out; the surrounding rocks melted and the meteorite was vaporised, filling the Earth's atmosphere with a cloud of dense dust. An unstable initial crater was formed, the sides of which then collapsed inwards, forming a final crater with a larger diameter, spanning the distance between what are now Johannesburg and Welkom. This was partly filled with impact melt that took millions of years to cool. In response to the impact, the centre of the crater began to rebound to form a wide, dome-like central uplift structure, upending the sedimentary formations to form the Collar Zone and bringing up granite-gneiss from great depths to form the Core Zone.

Today, after more than 2,000 million years of crustal uplift and removal of several kilometres of land surface by erosion, the vast crater has disappeared. What remains is the remnant of the central uplift and the geological ring structure that we call the Vredefort Dome.

Figure 19: Indication of size of Vredefort meteorite.

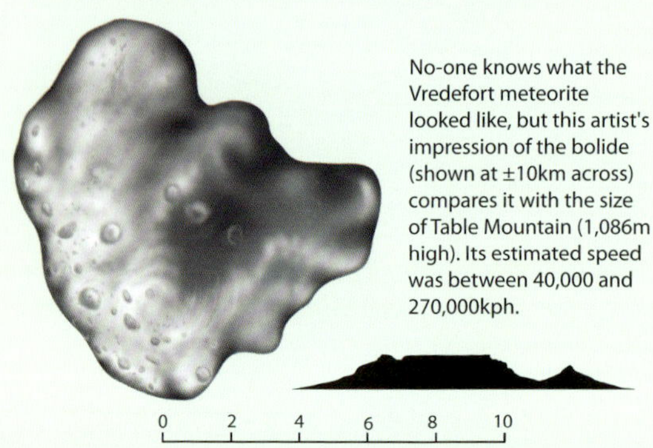

Figure 20: The likely effect of the meteorite impact and how the Vredefort Dome was formed.

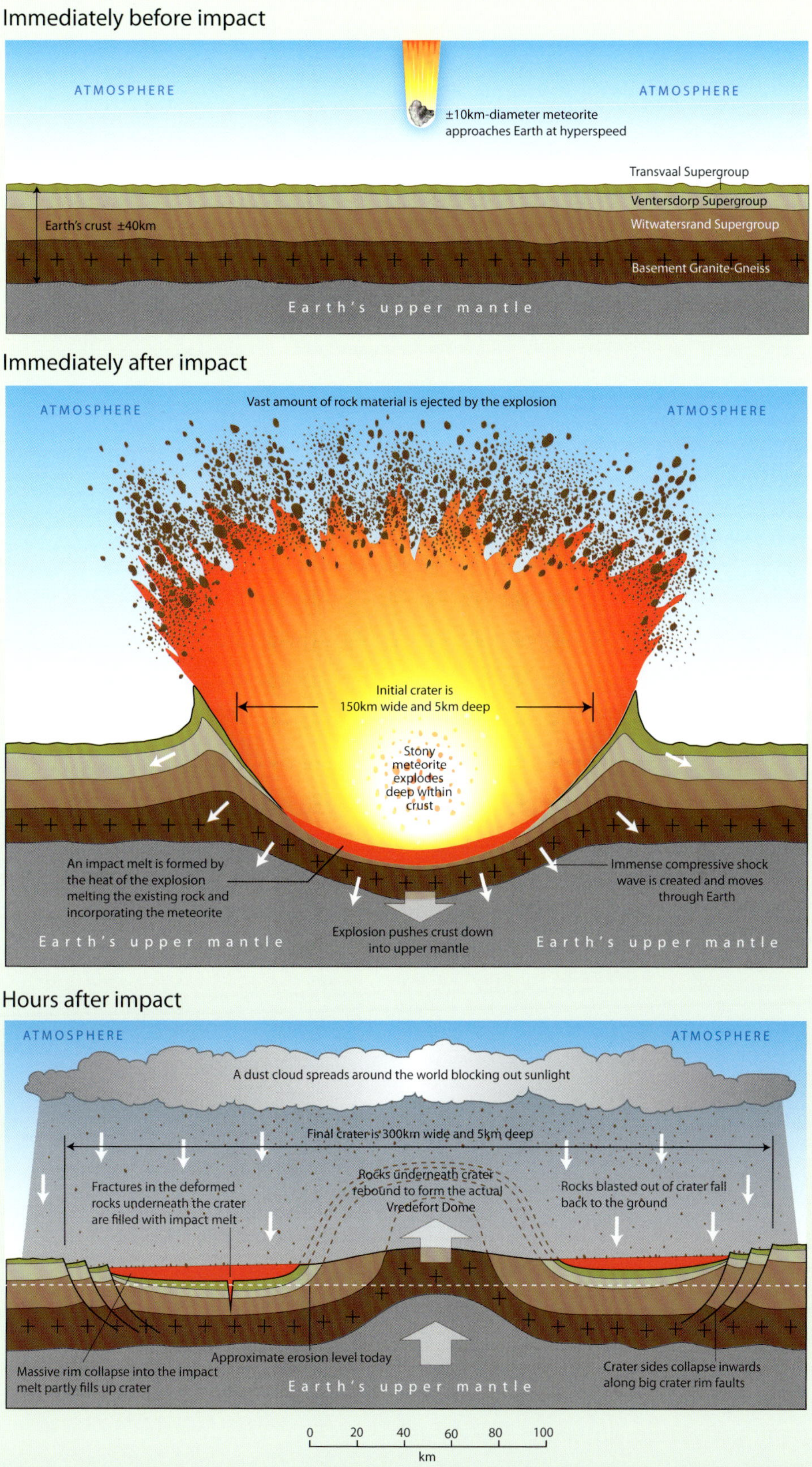

Adapted with permission from publicity pamphlet created by Prof. Roger Gibson (University of the Witwatersrand)

Things to see and do

The Vredefort Dome (see **Figure 21** opposite) provides a number of widespread must-see sites that tell the story of the impact event, and which provide irrefutable evidence for this Earth-shattering catastrophe.

- **Go to the Interpretation Centre (A):** Located about 3km north of the town of Vredefort off the R59, the Centre is situated on a bare koppie of Parys granite-gneiss and provides excellent all-round views of the Dome. Although it remains unfinished at the time of writing, the plans include up-to-date audio-visual and graphic information on the Vredefort Dome. It's a good place to start and from which to work outwards (site location numbers are indicated on the map; some are on private land so access is not guaranteed).
- **Vredefort granophyre (B):** On private land 1km east of Vredefort on the southern side of the Greenlands gravel road, narrow, dyke-like bodies cut through the granite-gneiss. In several places these dykes have provided a field canvas for petroglyphs (rock engravings) of San origin.
- **Pseudotachylite breccia (C, D):** Several quarries show this amazing breccia but the most accessible exposures are at the old Salvamento Quarry (**D**), or the nearby old Leeukop Quarry (**C**) (one of the satellite World Heritage Sites), both northwest of Parys.
- **Witwatersrand Formations (E, G):** There are good views along the Venterskroon gravel road (**E**), which follows the Vaal River. After the discovery of the Witwatersrand goldfield it did not take long for prospecting to start on similar conglomerate reefs in this area. By 1887 there was considerable activity in what became known as the Venterskroon goldfield but it never yielded much gold – the total reported production was only 123kg. The Amazon Reef (**G**), correlated with the Kimberley Reefs, is close to the Potchefstroom road about 6km northwest of Venterskroon; there is an interesting geological display at the small Venterskroon information office.
- **Shatter cones (F, H):** The best-known outcrop is in Maraisburg Formation quartzite at the western side of the Schoemansdrift bridge across the Vaal River, adjacent to the gravel road (**H**). An alternative site within fractured Booysens Formation shale is found about 2km north of Venterskroon in a road cutting (**F**).

The Interpretation Centre located on a Parys granite-gneiss dome popularly known as Die Koepel a few kilometres north of Vredefort. At the time of writing the facility was not yet open to the public.

The Collar Zone consists largely of the Witwatersrand Supergroup strata, which were forced up several kilometres by the rebound effect after the impact. The photo shows some of these geologically upturned hills near Parys.

Figure 21: The simplified geology of the Vredefort Dome showing major roads and location of selected sites.

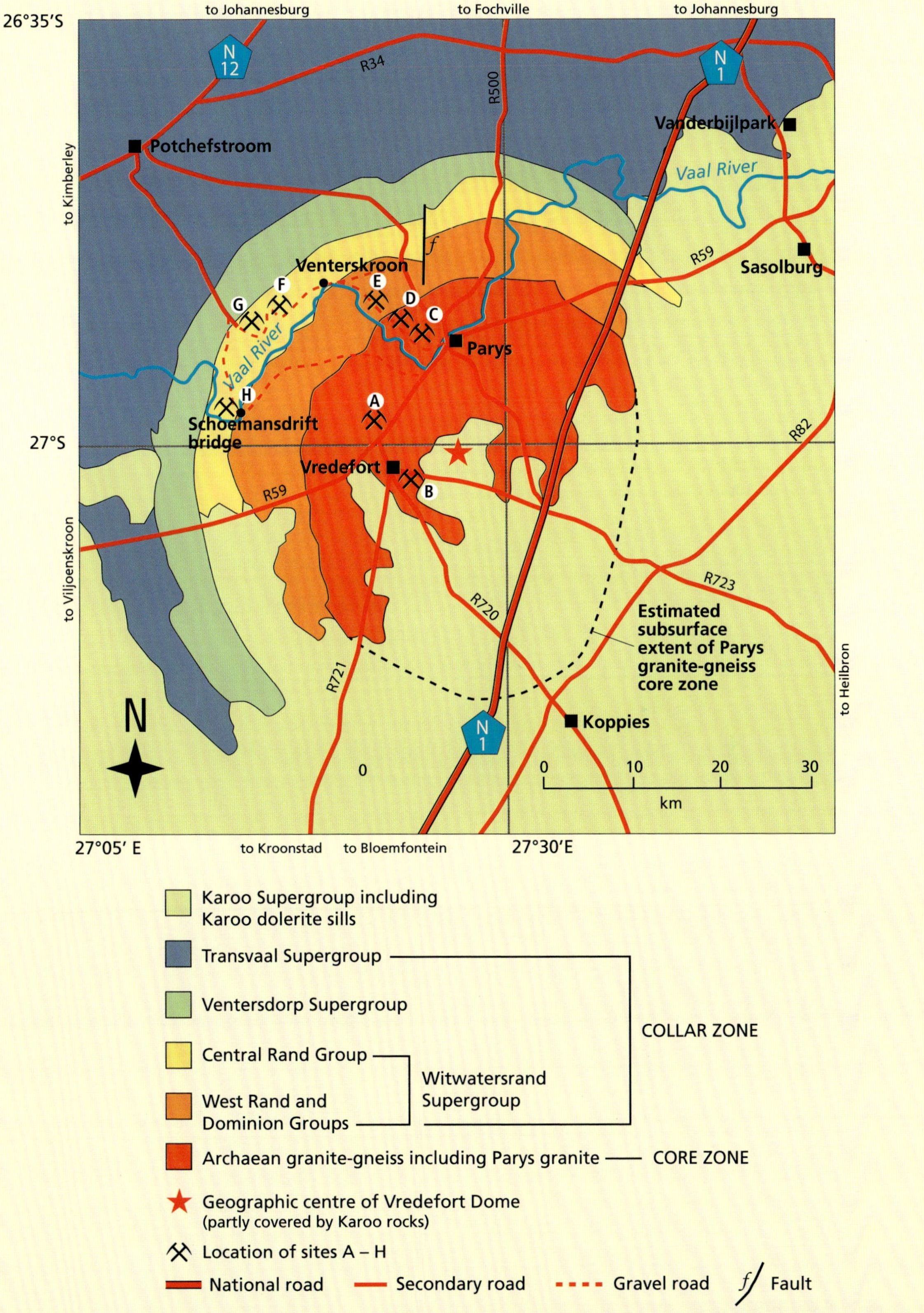

MPUMALANGA

OVERVIEW OF LANDSCAPES AND GEOLOGY

Mpumalanga is a geologically diverse region that is host to the famous Barberton Mountain Land and a portion of the eastern Bushveld Complex. The regional landscape can be divided into four major geomorphic domains, which correspond closely to the major geological formations of the region: the plains of the Eastern Lowveld; the strongly dissected Great Eastern Escarpment; the Mpumalanga Highlands; and Barberton's unique Makhonjwa Mountains. Geosites 24–31 are shown on the national maps (pp.12–13), and on the regional landscape and geological maps (pp.164–165).

The major geological formations of the region are listed below:

Barberton Supergroup
(Early to Middle Archaean)

This complexly deformed greenstone sequence is the largest in southern Africa, and is world renowned for its fine preservation of ancient geological processes and events, and signs of very primitive life that appeared more than 3,000 million years ago. These formations make up the undulating terrain of the Barberton Mountain Land, centred on the old gold-mining boom town of Barberton. The Barberton Makhonjwa Geotrail (**Geosite 24**) is a unique and expertly designed geotourism attraction that was declared a World Heritage Site in 2018.

The **Barberton Supergroup** is divided into three geological groups based on very different rock types and diverse environments of deposition. These are, from oldest to youngest:

- **Onverwacht Group:** This consists of a thick sequence of mainly mafic and ultramafic lavas (including the first recognised komatiite), pillow lavas that erupted under water, and lesser tuffs and cherts. Around 3,500 million years ago this was the start of an early mini-continent.
- **Fig Tree Group:** Consisting mostly of fine-grained sediments deposited slowly in a deep ocean under generally stable conditions, it includes colourful banded cherts, dark red-brown iron formations, shales and tuffs.
- **Moodies Group:** This comprises a variety of mainly shallow-water sedimentary formations deposited under alluvial, fluvial and shallow tidal marine conditions. The sediment was derived largely from granitic rocks that had intruded the mini-continent.

Basement Granite-Gneiss
(Archaean)

Basement granite-gneiss makes up most of the extensive Eastern Lowveld. It comprises a heterogeneous group of granites, gneisses and migmatites of varying Archaean age that weather and erode relatively easily, and are typified by the landscape of the Kruger National Park and the lower terrain around Nelspruit and the Barberton Mountain Land. These rocks are mostly slightly younger than the greenstones, and extend northwards into Limpopo. (No geosites relating to this formation are featured here.)

Transvaal Supergroup
(Late Archaean to Early Proterozoic)

Around 2,650 million years ago, rifting and stretching of the Kaapvaal Craton, followed by subsidence, resulted in the development of the Transvaal Basin, a major elongated depository, which can now be traced for over 500km east-west and some 200km north-south. This basin filled with the large, shallow Transvaal Sea and sedimentation on the submerged continent took place under fairly stable conditions until around 2,100 million years ago. Some of the geological formations that were produced had a profound effect on the Earth's environment and in later times produced spectacular landscapes, as mentioned below.

- **Wolkberg Group:** This sequence was the first to form, deposited within a smaller proto-basin, and it is exceptionally well showcased in the northeast, along the Great Eastern Escarpment. These mostly sedimentary formations are displayed under a blanket of Black Reef quartzite at God's Window (**Geosite 29**), Bourke's Luck Potholes (**Geosite 30**) and Blyde River Canyon (**Geosite 31**).
- **Black Reef Formation:** This thin, hard quartzite formation, almost horizontal in attitude, blankets the landscape of the Great Eastern Escarpment. It was deposited on the floor of the inland Transvaal Sea as a clean, pebbly sandstone less than 2,650 million years ago. It formed a covering that long resisted strong erosion, and protected the underlying softer Basement granite-gneiss and Wolkberg rocks. There is no better place to see it than at Kaapsehoop (**Geosite 25**).
- **Malmani Subgroup** (lower part of **Chuniespoort Group**)**:** Following deposition of the Black Reef sandstones, a period of very stable continental platform conditions persisted within the Transvaal Basin for around 150 million years. In a warm and shallow sea, vast beds of blue-green algal life flourished on a scale never before seen on Earth (see also p.184). The geological end-result is the extensive beds of stromatolitic Malmani dolomite, deposited from around 2,500 million years ago, as seen today at Sudwala Caves (**Geosite 27**) and Lone Creek Falls (**Geosite 28**).
- **Pretoria Group:** After the Malmani dolomite had become exposed hard rock, a long period of weathering and erosion followed, resulting in the formation of a palaeo-surface. On this ancient surface a thick deposit of almost insoluble chert fragments formed – a residue from thin chert layers within the dolostone – resulting in a very resistant, chert-rich unit known as the Bevets conglomerate. This is now the cause of several waterfalls, such as Lone Creek Falls (**Geosite 28**). Thereafter, various shales and sandstones, plus inter-bedded mafic lavas, slowly accumulated in the Transvaal Basin, making up a package about 6km thick, as seen in the Mpumalanga Highlands. At Waterval Boven one of these units, the approximately 100m-thick Daspoort quartzite, forms a hard barrier and knickpoint on the Elands River (**Geosite 26**).

Bushveld Complex
(Early Proterozoic)

The **Bushveld Complex** is the world's largest layered mafic intrusion by far, and contains the greatest concentration of mineral wealth on Earth. It is enormous and astonishing: 440km from east to west, more than 250km from south to north, and made up of several distinct 'limbs'. It is best exposed in the hilly Eastern Limb to the west of the Mpumalanga Highlands. Here, there are exceptional exposures of approximately 9km-thick layered igneous formations, which are divided into several different rock-type zones that extend for hundreds of kilometres; this is demonstrated at the enigmatic Dwars River Chromite exposure (in the section on Limpopo, **Geosite 32**) of layered anorthosite and chromitite – a feature still not fully understood.

The Great Eastern Escarpment

This is the most spectacular part of the region and the landscape calls for explanation. It rises steeply from the Eastern Lowveld to form a relatively narrow mountainous zone, deeply incised by river valleys, and leading westwards onto the Mpumalanga Highlands. Formation of the escarpment is related to a combination of geological and geomorphological phenomena that began around 150 million years ago along the eastern coastal margin: crustal rifting preceding the break-up of Gondwana, uplift of the subcontinent, and denudation on a massive scale. During the Cretaceous, the wetter tropical climate created the ideal situation for weathering, erosion and removal of material by many youthful, eastward-flowing rivers that cut back into the newly formed coastal scarp. The result was the geologically rapid retreat of the scarp across the coastal plain, a forerunner of today's Great Eastern Escarpment. The earlier coverings of Karoo and Transvaal sequences have long since been removed by erosion.

Figure 22: Landscape map of the Mpumalanga region. This shows admirably the relationship between landscape and geology. The high ground is shown in tones of brown and yellow, ending at the partly dissected Great Eastern Escarpment, while the Eastern Lowveld is shown in blueish-grey. The Barberton Mountain Land (also known as the Makhonjwa Mountains) appears in the southeast, extending into Swaziland.

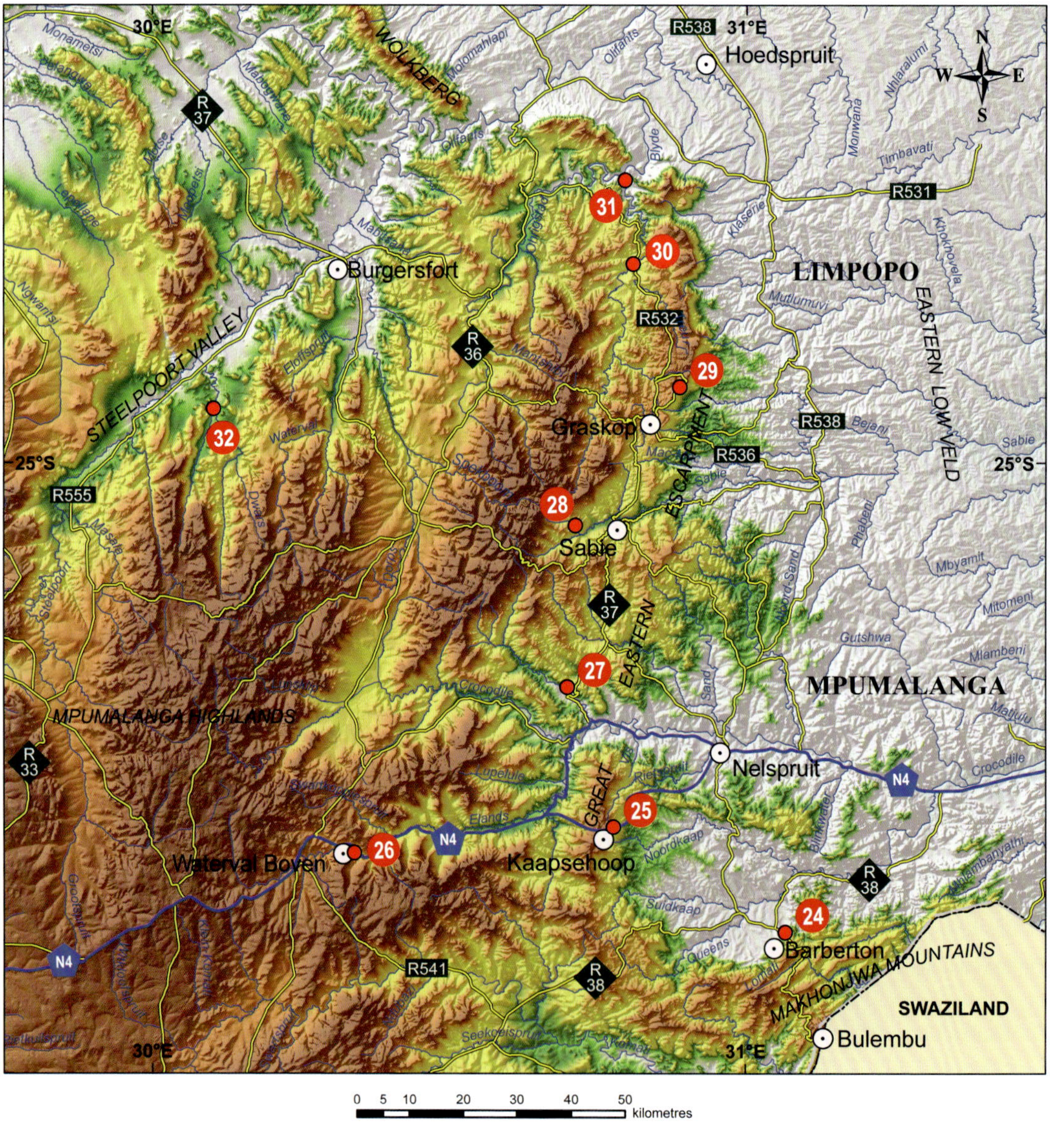

Figure 23: Geological map of the Mpumalanga region. In the west the geology is dominated by the Transvaal Supergroup, intruded by the Eastern Limb of the Bushveld Complex. In the east the granite-gneiss of the Basement underlies the Eastern Lowveld, while in the southeast there are Archaean greenstones of the famous Barberton Supergroup (also known as the Barberton Greenstone Belt).

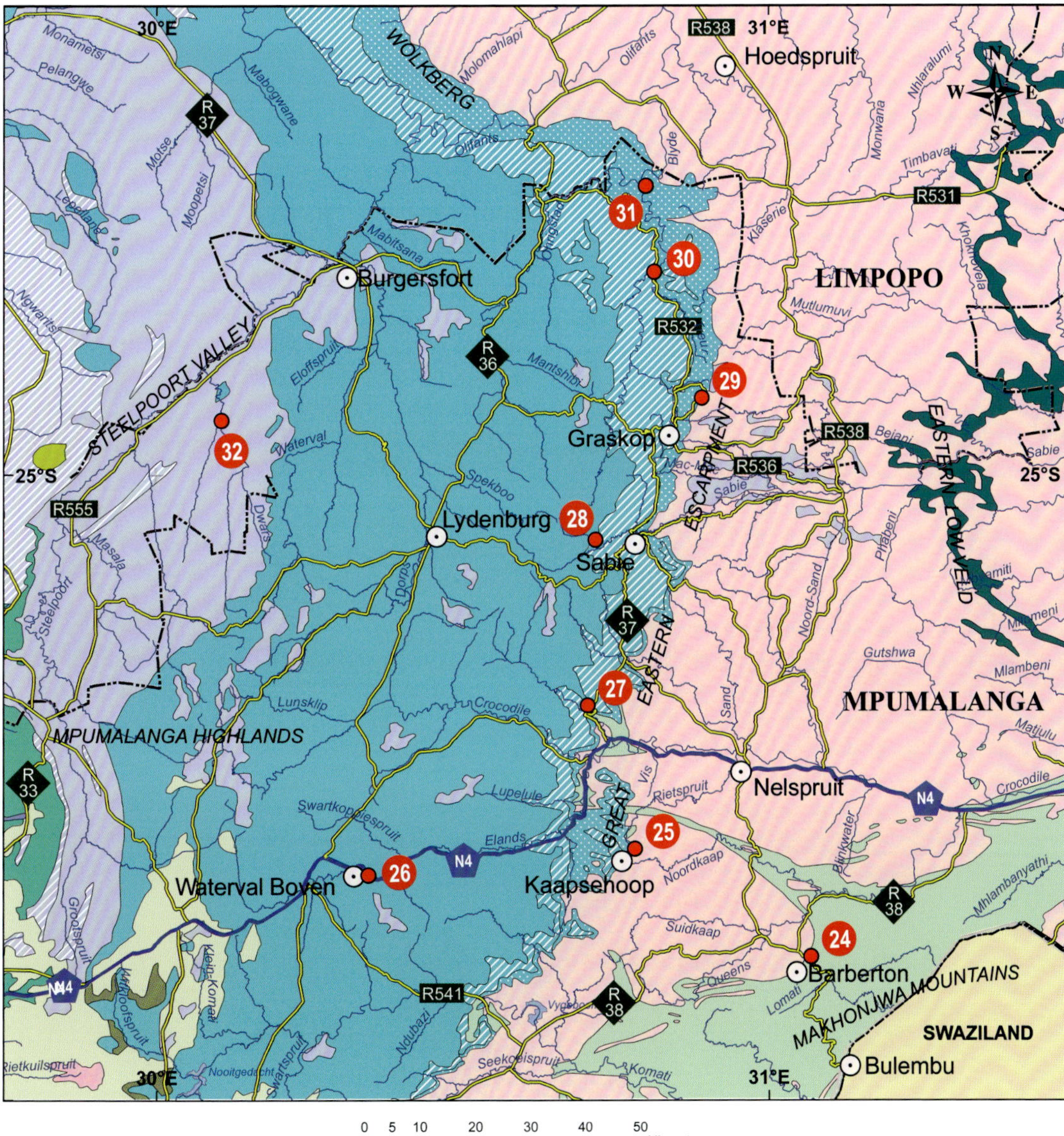

• A stratigraphic key to this simplified regional geological map appears on the inside covers

24

Barberton Makhonjwa Geotrail

WORLD HERITAGE SITE

A fascinating road trail that reveals some of the oldest and best preserved geology on Earth

The northern part of the Barberton Mountain Land seen from site BM7 on the new and extraordinary Barberton Makhonjwa Geotrail. Known as Makhonjwa/Lebombo View, this superbly panoramic location also displays fine specimens and offers interpretations (including in Braille) of all rocks seen at geosites along the geotrail.

The Barberton Mountain Land is a world-famous geological terrain and is the most intensely studied area of Archaean geology on Earth. These rock formations show the diverse geological and life-forming processes that were taking place 3,600–3,100 million years ago as one of the Earth's earliest micro-continents was being formed, which would eventually become part of the Kaapvaal Craton.

Location and getting there

- Barberton is situated at the western foot of the Makhonjwa Mountains, east of the Great Escarpment and close to the Swaziland border. Several roads lead into Barberton: the R38 from Badplaas or from Kaapmuiden, and the R40 from Nelspruit. The geotrail is a continuation of the R40 to the Swaziland border at Bulembu, which has recently been reconstructed and tarred.
- GPS (site BM1 on geotrail): 25°46′13″S; 31°03′57″E

Setting the scene

The Makhonjwa Mountains (also called the Barberton Mountain Land) give their name to the geological road trail along which some of the oldest rocks on Earth can be seen and interpreted, a unique facility implemented in anticipation of achieving coveted World Heritage Site status as the most outstanding and best preserved Archaean geological terrain on Earth. The route follows a scenic, 37km-long, undulating and winding mountain road between Barberton and the border at Bulembu, and presently comprises 11 demarcated and described geosites, including spectacular viewpoints. At each geosite there is parking and a vandal-proof panel presenting geological information and concepts at both scientific and easy-to-read levels.

This is ripped-up cherty breccia from site BM8, named Tsunami Conglomerate. It is thought to have formed from some rapid, catastrophic event, such as a tsunami.

A typical view of the Barberton Mountain Land looking southwards from site BM9, called Banded Ironstone. The area is underlain by banded iron formation of the Fig Tree Group.

Figure 24: Locality map for the Barberton Makhonjwa Geotrail indicating named sites (BM1–11) and distances from the start.

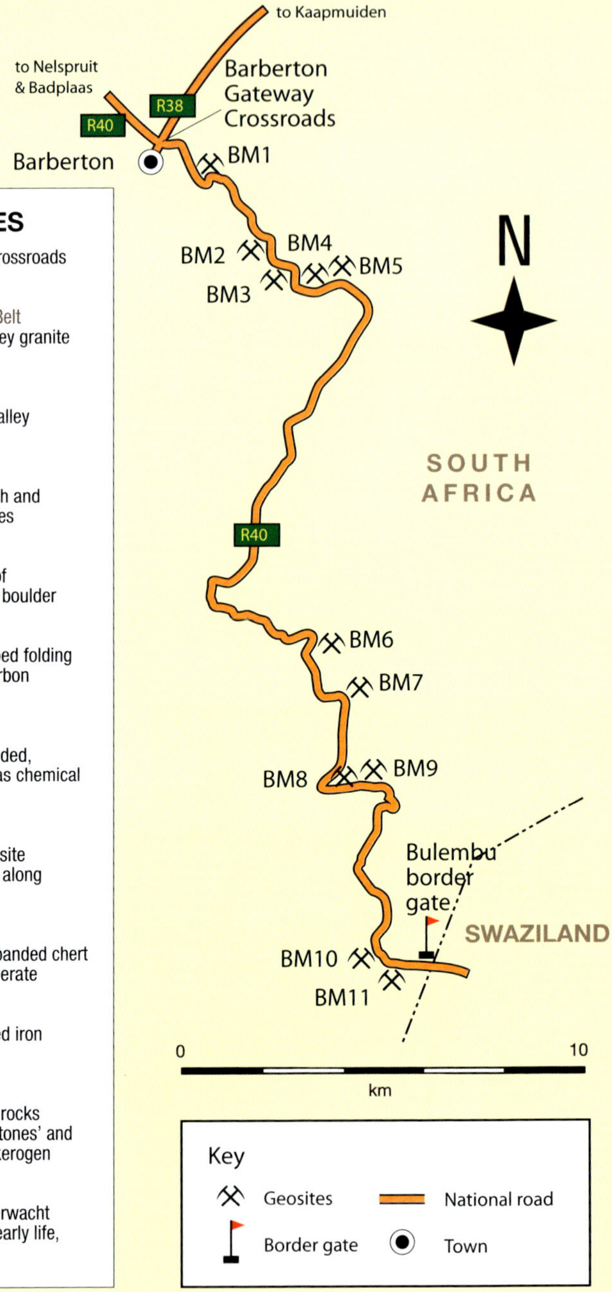

SELECTED GEOSITES

Distances are from the Barberton Gateway Crossroads

BM1 (2.4km)
Greenstone View and Barberton Greenstone Belt Boundary. Faulted contact between Kaap Valley granite and Onverwacht greenstone

BM2 (6.3km)
De Kaap Valley View. Best view of the Kaap Valley tonalite pluton

BM3 (7.0km)
Tidal Sandstone. Moodies sandstone, clay-rich and showing shallow-water sedimentary structures

BM4 (7.3km)
Eureka View and Alluvial Conglomerate. Top of Saddleback Pass and Moodies conglomerate boulder

BM5 (8.2km)
Dycedale Syncline and Biomats. Tight U-shaped folding in Moodies and highly compacted organic carbon representing early microbial life

BM6 (23.6km)
Painted Quarry and Baryte. Tightly folded banded, coloured, ferruginous Fig Tree chert, formed as chemical precipitate in deep water

BM7 (25.3km)
Makhonjwa/Lebombo View. Spectacular viewsite and specimens of rocks occurring elsewhere along the geotrail

BM8 (27.7km)
Tsunami Conglomerate. Fig Tree pink shale, banded chert and reddish iron oxide in an angular conglomerate

BM9 (29.2km)
Banded Ironstone. Jasper-rich Fig Tree banded iron formation with minor chert and volcanic tuff

BM10 (36.3km)
Volcanic Lapilli and Black Chert. Onverwacht rocks showing grey chert with small volcanic 'hailstones' and also banded black chert that contains fossil kerogen

BM11 (36.6km)
Pillow Lavas and Msauli/Bulembu View. Onverwacht chert that contains microscopic evidence of early life, and nearby, submarine lava flows

Regional geology

This is the unrivalled Barberton Greenstone Belt, 140km long and up to 40km wide, an exceptionally well-preserved, thick and varied sequence of Archaean rock formations. They are strongly faulted and tightly folded into steep synclines and anticlines, resulting in a high degree of structural complexity, and entirely surrounded by large batholiths and plutons of Archaean granitic rocks of different ages. The full geological sequence that makes up the **Barberton Supergroup** comprises, from oldest to youngest:

● **Onverwacht Group** (3,570–3,260 million years ago) – comprising mainly ultramafic (komatiite) and basaltic lavas in various forms; these are the Earth's earliest recognisable volcanic rock formations, which formed as island arcs.

● **Fig Tree Group** (3,255–3,225 million years ago) – mainly sedimentary rocks such as impure shale and sandstone derived from eroding island arcs and volcanic activity, and sediments such as banded ironstone and cherts that were deposited in deep water.

● **Moodies Group** (3,225–3,215 million years ago) – mainly coarser sandstones and conglomerates and minor shales, deposited under shallow-water conditions, such as tidal shorelines and deltas, as the mini-craton evolved and stabilised.

Finely banded and tightly folded iron-rich chert make a colourful display at site BM6, known as Painted Quarry.

At site BM10, called Black Chert, there is very finely banded grey to blackish Msauli chert, part of the Onverwacht Group.

At site BM4 (Eureka View) a large boulder of Moodies Group conglomerate has moved about 100m down the steep hill. Most of its cherty pebbles originally came from the Onverwacht Group, plus some Fig Tree Group volcanic rocks.

Landscape and landforms

The landscape is one of strongly dissected and steeply undulating topography, showing great natural beauty and splendid scenery. From Barberton, at an elevation of around 850m, the road climbs to 1,360m at Saddleback Pass and the mountain reaches 1,640m at Saddleback Hill. The varied geology, the undulating terrain and the generally high rainfall have resulted in considerable biodiversity, so much so that the region is a biological hotspot. Land is now being used for timber plantations and several nature reserves have been established. As can be expected from a historic goldfield, there are numerous scars from earlier mining activity.

Rock formations

● There are hundreds of places of geological interest in the Barberton Mountain Land and it is a favourite destination for ongoing geological research and academic visits. The Barberton Makhonjwa Geotrail enables observation of a variety of these rocks in the field from a representative sequence of Moodies, Fig Tree and Onverwacht Groups, revealed at the marked geosites as well as at several undeveloped sites.

● The informative guidebook to the geotrail is essential reading. The map (opposite) provides the location and name of each of the documented geosites.

The Barberton gold rush

The rush for gold at Barberton is a legendary event in South African history. In 1882 alluvial gold was found in the De Kaap Valley, and some diggers moved on to hills on Moodies Estate. There Auguste Robert ('French Bob') found alluvial gold in Concession Creek and in June 1883 he traced it to the Pioneer Reef – the first payable gold reef in the Barberton area. Soon at least 1,000 diggers had joined the gold rush, working on their claims and living in hastily set-up camps. In the same hills further to the northeast, the Barber and Rimer brothers found more rich gold reefs and this led to the establishment of boom-town Barberton in June 1884. But it was Edwin Bray's discovery of the fabulous Golden Quarry deposit alongside Sheba Creek in May 1885 that led to the Barberton goldfield becoming known worldwide. Up until then diggers had been looking for easily won alluvial gold, but new methods were now required, involving blasting and crushing the hard ore to release the precious metal. More than 350 gold-mining operations and prospects were eventually recorded but most production has come from five mines: Sheba, Fairview, New Consort, Agnes and Barbrook, all still operating nearly 130 years after the first discovery of gold, although in 1886 the Barberton mines were eclipsed by the new Witwatersrand goldfield. The estimated total gold production from the Barberton mines is about 345 tons, or over 11 million ounces.

Things to see and do

- **Drive the Barberton Makhonjwa Geotrail:** The full 37km-long road could take you a day, or it can be done in part only. The signposted and numbered geosites and viewpoints have geological information revealing the significance of each place. Ideally, you need the official 2013 guidebook entitled *Barberton Makhonjwa Geotrail* (available from Barberton Tourism), which provides considerable information on each site.

Part of the valley at Msauli View as seen from site BM11, and the very illustrative and explanatory signage.

Welcome sign at the Barberton Gateway Crossroads (R38/R40) and the start of the geotrail.

- **Visit the Golden Quarry and the site of old Eureka City:** These are situated on the property of Sheba Mine and a fascinating visit can be arranged through Barberton Tourism. The Golden Quarry is reputed to be the richest gold deposit ever mined. A colourful mining camp that became a small town called Eureka City was established in 1885 on higher ground above Sheba Creek to avoid the heat and malaria, but no longer exists.

25 Kaapsehoop Escarpment

A HISTORIC GOLDFIELD

Rugged terrain, weirdly eroded rock formations and ghostly swirling mists led to this once being known as Duiwels Kantoor

An eroded gap in Black Reef Formation quartzite on the edge of the Kaapsehoop Escarpment, seen as one faces northwards. Timber plantations mostly cover the Kaap Valley tonalite (a variety of granite) below, and the hills in the middle distance are the extreme western end of the Jamestown Schist Belt.

Once the scene of frenetic gold prospecting, the secluded village of Kaapsehoop now caters for passing travellers and harbours nothing more threatening than herds of wild horses. In 1882, struck by mountain scenery reminiscent of the Cape, then-President Martinus W Prinsloo optimistically renamed the place Kaapsche Hoop and the broad valley at the foot of the escarpment became the De Kaap Vallei.

■ Location and getting there

- The hamlet of Kaapsehoop, as it is now spelt, lies on the narrow Godwan Plateau, an almost isolated, spur-like remnant of Highveld preserved between the Elands River Valley and the escarpment. You can reach the village either by turning off the N4 at Ngodwana and going a further 12km; or from Nelspruit, turn off the N4 immediately outside the town and take a very scenic drive of about 30km up the escarpment.
- GPS: Kaapsehoop (escarpment viewsite): 25°35′39″S; 30°46′44″E

Setting the scene

This is the original stone-built Mining Commissioner's office in Kaapsehoop. Dating from 1884, it was built to control gold claims in the area, and was the most substantial building in the village.

In 1881, with the granting of a mining concession for the Pilgrim's Rest Gold Field, many of the itinerant prospectors moved away, heading south along the Great Eastern Escarpment where small gold finds continued to be made. Along the edge of the Godwan Plateau alluvial gold was discovered in creeks on the farm Berlyn (now Berlin) in early 1882. The area was soon proclaimed a public digging and, on nearby state land at the very edge of the escarpment, a bustling mining camp quickly grew at a place known as *Duiwels Kantoor* or *Kantoorberg*. It became such a magnet for prospectors that a local Gold Commissioner's office was soon created.

The Kaapsehoop Escarpment, seen looking southwards. The distant flat land surface is underlain by a relatively thin layer of resistant Black Reef quartzite, seen outcropping on the right as a multitude of small rock pinnacles and pillars.

Regional geology

The weirdly shaped rock formations are weathered exposures of the **Black Reef Formation**, a widespread sheet sandstone (now quartzite) and conglomerate formation that was deposited on the floor of the developing Transvaal Basin about 2,650 million years ago. Here at the escarpment edge it overlies either Archaean **Kaap Valley tonalite** (a granite), or it was deposited over the small Godwan Basin, a precursor to the Transvaal Basin. The name 'Black Reef' is an old mining term for this sometimes dark and occasionally gold-bearing formation.

The Black Reef sandstone (now quartzite) was deposited in shallow water. This remnant outcrop shows both angular cross-bedding (towards the top) and parallel (planar) bedding (lower down) indicating sedimentation under differing water-flow conditions.

Landscape and landforms

It is the hard and almost horizontal Black Reef quartzite that forms the top of the impressive Great Eastern Escarpment and the extraordinary eroded landscape (see Mpumalanga Overview, pp.162–163). But there is more to it – a Google Earth image shows a close-spaced pattern of structural rock joints more or less at right angles, imparting a strong crisscross effect on the quartzite layer. This high-rainfall mountain edge provides an ideal pathway for moisture that penetrates existing cracks or joints in the rock; there, it may freeze, exacerbating the cracks in a process known as 'freeze-thaw weathering' – so accelerating erosion. The depth of at least some of the apparently natural gulleys is probably the result of soil excavation during gold mining.

Gold mining at Kaapsehoop

Alluvial gold was found near here in 1882, in a creek on the nearby farm Berlyn, and was traced to a reef associated with a diorite dyke. Soon many prospectors were busy in the area, and a new rush of fortune-hunters started. Generally, the gold found was the product of rock erosion (eluvial), found in soil and washed from rubble zones across the very irregular land surface. It probably came from weathered quartz veins or 'rotten reefs' similar to the Pilgrim's Rest deposits. Alluvial gold was also found in streams, but because of the very rugged terrain, concentration of the sought-after metal was poor. In spite of some rich alluvial deposits and the fairly common occurrence of gold nuggets, by 1883 the easy-found gold had run out and many of the prospectors had moved down into the fever-ridden De Kaap Valley. Then everything changed in June 1883, when Auguste 'French Bob' Robert found the Pioneer gold reef at Moodie's Estates (see 'The Barberton gold rush', p.170) near Barberton, where the many rich reefs soon eclipsed Kaapsehoop's rather meagre gold.

Nevertheless, some surprising gold nuggets were found in the Kaapsehoop area. In 1882 the largest nugget found weighed nearly 10 ounces and in 1883 this was presented by the South African Republic to the King of The Netherlands for his support during the 1st Anglo-Boer War. In 1912 the amazing Peacock Nugget weighing a massive 179.8 ounces was found 1km south of Kaapsehoop, and was sent to England, where a replica is displayed in the British Museum. Much later – and well after the time of the early gold rush – gold was mined by way of numerous adits into the basal conglomerate, gold-rich in places but erratic. Today there is very little evidence of any old mining.

Photographer unknown. From the collection of Carl Anhaeusser

This undated old photo, perhaps taken around the early 1900s, shows hardrock mining taking place on the edge of the Kaapsehoop Escarpment.

Kaapsehoop is now well known for its herds of wild horses of uncertain origin. They wander freely through the village and grassy outcrops, and are a reminder of times gone by. Part of the eroded Black Reef Formation quartzite landscape is seen in the background.

Rock formations

- At Kaapsehoop the Black Reef Formation is about 30m thick and comprises a zone of basal conglomerate and a thicker sequence of clean, well-bedded quartzite, including some thin shaley and pebbly layers.
- A good look at the exposed rock pillars shows planar bedding and cross-bedding, indicating the deposition of clean quartz sand and pebbles on a river flood plain or in shallow water along the margin of the early Transvaal Sea.

Things to see and do

- **See the view:** There is nothing better than to be on the edge of the 1,700m-high Kaapsehoop Escarpment, ideally early in the day, for the incomparable view. The verdant Kaap Valley, some 1,000m below, is often covered in a surreal sea of morning mist.
- **See the regional geology:** To the northeast are the rounded hills of Nelspruit's Archaean granite-gneiss terrain; to the east are the hills of the Jamestown Schist Belt; far in the distance to the southeast lies the Barberton Mountain Land; and the wide valley below is made of intrusive Kaap Valley tonalitic granite.
- **Walk through the labyrinth:** Look for bedding structures and pebbly conglomerates in the weathered quartzite maze and appreciate the eroding power of nature, particularly in relation to rock joints.
- **Drive Kaapsehoop Pass:** The road to Nelspruit takes you down to the Kaap Valley and you will cross the narrow Jamestown Schist Belt. Look out for the old serpentine asbestos (chrysotile) mine and for veins of rare stichtite, a soft, mauve-purple, talc-like mineral found in a road cutting through dark greenstone.

26

Waterval Boven

PROVINCIAL HERITAGE SITE

Once a formidable geological barrier along the railway line to the coast, this is now the rock-climbing centre of South Africa

The Elands River as it plunges about 75m over the falls of the same name, seen from near the old NZASM tunnel. This knickpoint on the river effectively separates the bushveld of the Eastern Lowveld from the grasslands of the Highveld. Notice the strong rectangular jointing in the wall of resistant quartzite.

One of South Africa's foremost historic railway towns, Waterval Boven was established as a rail depot for the Nederlandsche Zuid-Afrikaansche Spoorwegmaatschappij (NZASM) on the edge of the escarpment in about 1893. The town's new official name is eMgwenya, meaning 'place of the crocodile' in siSwati or isiZulu.

Location and getting there

- Formerly known as Waterval Boven (meaning 'above the waterfall' in Dutch) the town is located adjacent to the N4, about 250km from Johannesburg and 110km from Nelspruit. The highway skirts the northern side of town, passes through a road tunnel and follows the scenic valley of the east-flowing Elands River down to the Lowveld. The impressive Elands waterfall is close to the western end of the old railway tunnel but not easily seen from the main road.

- GPS (tunnel parking area, off the N4): 25°38′22″S; 30°20′46″E

Setting the scene

Construction of the legendary NZASM railway (see text box on p.179) was a monumental task: one of the biggest challenges was getting up the escarpment, here deeply incised by the Elands River Valley. Between the two old towns once known as Waterval Onder and Waterval Boven, a thick horizontal layer of reddish, cliff-forming rock on both sides of the valley effectively bars the way, and the only solution was to build a tunnel through this barrier. Declared a National Monument in 1962, this historic 211m-long, single-line railway tunnel was first used in 1895 but was replaced in 1908, when the railway took another, less steep route on the other side of the river and the old railway route became the main road. When the new, 400m-long, double-lane road tunnel on the N4 was opened in 1972, the old tunnel was abandoned.

Regional geology

This part of Mpumalanga is underlain by the **Pretoria Group**, the mainly sedimentary and upper part of the **Transvaal Supergroup** that filled the enormous Transvaal Basin 2,650–2,200 million years ago (see Mpumalanga Overview, pp.162–163). Here, the main barrier is the hard, cliff-forming quartzite of the **Daspoort Formation**, less than 100m thick. Much of the surrounding countryside is also underlain by diabase sills that intruded the almost horizontal sedimentary rocks at various levels, and reflect an early phase of the **Bushveld Complex**.

Landscape and landforms

This part of the Great Eastern Escarpment forms a broad zone of strong headward

An outcrop of the resistant Daspoort quartzite close to the falls. It is this hard rock that has formed the knickpoint on the Elands River and created the spectacular falls.

The eastern entrance to the 1972 road tunnel made through resistant blocky quartzite. There is parking, and local guides can escort you to the old NZASM tunnel.

An aerial view of the Elands River Falls and part of eMgwenya. Notice that on the left-hand side the N4 highway disappears into the quartzite hillside and re-emerges about 400m later. The small road in the foreground is the original path of the NZASM railway line before it enters the old tunnel.

erosion – the action of a river extending its valley upstream – characterised by scarps and deeply incising, boulder-filled valleys. Various faults have certainly enhanced the progress of the eroding rivers. The Crocodile River and its main tributary, the Elands River, have successfully cut back through the lower formations and are now actively working into the upper formations and diabase sills of the Pretoria Group. The Elands River waterfall marks the position of an important knickpoint on the escarpment, essentially where the Lowveld rapidly changes to Highveld.

The NZASM Railway Tunnel

The ambitious Nederlandsche Zuid-Afrikaansche Spoorwegmaatschappij (NZASM) railway line started at Komatipoort in 1887 but was officially opened only in 1895 when it reached Pretoria. It ran from Delagoa Bay (now Maputo) to the capital of the old South African Republic, and provided the shortest route to the sea, while also avoiding British territory. Between Waterval Onder and Waterval Boven the old line rose 228m, with a maximum gradient of 1:20. A key link was the construction of a 3km-long rack-railway system, driven by a cogged rail between the bearing rails, that effectively negotiated the tortuous climb. This was the only tunnel along the entire route, and its building was an exceptional feat of engineering, considering the time and place, and that it was built on a curve and with a slope. Work on the tunnel itself started in October 1892 and blasting through the rock was completed – without serious mishap – in under a year, the hard quartzite being ideal for tunnelling. For stability, the inside was lined with dressed stone and impressive portals were constructed. In spite of these sterling efforts, the line was never a great success and in 1908 it was re-routed on the opposite side of the river along a 14km-long, less arduous track, with two tunnels. With the construction of these new railway tunnels, the old tunnel was recycled for road traffic until a wider road tunnel was opened in 1972.

The dressed stonework entrance at the western end of the NZASM tunnel, which was completed in 1893. You need to enter the tunnel from the other side and walk through it to get here and see the falls.

The picturesque valley of the Elands River below the falls. The resistant layer of Daspoort quartzite is clearly seen, as is the tunnel entrance (upper centre) in the distance.

Rock formations

- In this craggy area the rock succession is dominated by horizontal layers of hard, resistant quartzite and numerous diabase sills, where the softer, shaley layers have eroded away more easily.
- The Daspoort Formation makes a prominent cliff-forming quartzite and was once a well-sorted sheet sandstone formation deposited and reworked in the shallow Transvaal Sea.
- Over many millions of years the sandstone became hardened to form quartzite, and an orthogonal system of rock jointing is strongly developed. This provides the smooth, vertical rock faces favoured by rock climbers.

Things to see and do

- **Walk through the NZASM tunnel:** An escorted walk will take you through the ~211m-long old tunnel and you can marvel at the exceptional construction, carried out over 120 years ago. Outside the tunnel have a look at the hard Daspoort quartzite formation.
- **See the Elands River Falls:** The river plunges an impressive 75m over a well-jointed quartzite cliff into a deep pool and then flows through a scenic gorge below. You may see rock climbers moving precariously up the adjacent rock faces. To get to the viewing platform you need to go through the old tunnel.
- **Rock climbing & abseiling:** eMgwenya has established itself as a premier adventure-sport venue. Well over 500 graded climbing routes have been mapped out along the extensive vertical-to gently overhanging crags in the area, formed by the vertical retreat of the hard, resistant quartzite formation. Here, both the climate and rock conditions are excellent for this sport.
- **Visit the Stone Circle Museum:** This fascinating venue in eMgwenya has been created by controversial new-age scientist and author Michael Tellinger. Scattered in this region are many circular stone ruins, interpreted by some as having been built by an ancient human race at the dawn of civilisation. If you have the time it is worthwhile going on a memorable Stone Circle tour.

Johan Heine, rights holder, Zulu Planet Publishers

One of the many intriguing 'stone circles' found in the region. These are interpreted by some as evidence of an ancient civilisation; the more accepted explanation is that they are the remains of many hundreds of stone-walled settlements of pre-colonial African (baKoni) people.

27 Sudwala Caves

AN UNDERGROUND GEOLOGICAL WONDER

These historic caves are among South Africa's best-known show-caves

Courtesy of Sudwala Caves

An interior view near the entrance to historic and spectacular Sudwala Caves, located in the valley of the Houtbosloop, a tributary of the Crocodile River, and once thought to be the hidden location of the 'Kruger millions'.

The caves are found within a rugged dolomite hillside known as Mankelexele (meaning 'crag-on-crag'). Prehistoric man *Homo habilis* probably occupied them around 1.8 million years ago. Numerous Early Stone Age tools attributed to *Homo erectus* have been found in the area. (See 'How limestone and dolomite caves are formed' on p.84.)

Location and getting there

- The caves are located on the farm Sudwalaskraal, hidden by dense woodland in the valley of the Houtbosloop, a tributary of the Crocodile River. To get there from the N4 take the R539 to Sabie near Montrose, about 35km west of Nelspruit, and follow the signs. The caves are about 8km north of the N4 turn-off and about 2.5km from the Sabie road.
- GPS (parking area): 25°22′11″S; 30°41′58″E

Setting the scene

In the 19th century the caves were a refuge for Swazi chiefs, including one known as Sidwaba or Sudwala, and later during the 2nd Anglo-Boer War, Boer forces used the caves to store ammunition. Local legend says the 'Kruger millions' (in gold sovereigns) were hidden in them. The fortune disappeared, allegedly somewhere between Waterval Onder and Nelspruit, but searches have always proved fruitless. Nevertheless, large quantities of valuable bat guano were dug out in 1914 for use as fertilizer. In 1965 the farm was purchased by PR Owen, who saw the tourism potential of the caves and opened them to the public.

The Sudwala Caves

A spectacularly large cavern, known as the PR Owen Hall, lies about 150m from the entrance. This amphitheatre-like cavern is 70m wide, has a dome-like roof up to 37m high, can seat more than 500 people and has been used as a concert venue. The caves are well ventilated by a constant stream of fresh air and the temperature remains at an even 17°C. The cave system is large, extending for many kilometres into the mountain, and is said to consist of more than 30km of passages. It takes a dog-leg course, selectively following a crisscross pattern of joints and at least one fault. The numerous cave formations have been given popular names. The caves are now extremely dry, at least partly due to the development of timber plantations on the adjacent higher ground.

Regional geology

The cave system has developed within the **Oaktree Formation**, the lowermost dolomite formation of the **Malmani Subgroup**, which lies immediately above the **Black Reef Formation** (see Kaapsehoop Escarpment, pp.171–175).

The caves are found in these rugged dolomite hills, located on the thickly wooded slopes on the upper right-hand side.

Courtesy of Sudwala Caves

The landscaped entrance to Sudwala Caves reveals the horizontal beds of stromatolitic Malmani dolomite (dolostone) into which the caves were eroded by karsting.

You can see evidence of very ancient algal life outside of the caves. This is a top view of a layer of small stromatolites, and evidence for the dolomite's biogenic origin.

An excellent exposure of eroded Malmani dolomite near the caves entrance. The almost horizontal bedding, abundant crinkly algal laminations and thin irregular layers of whitish chert are well shown.

The Malmani dolomite (or dolostone), deposited around 2,500 million years ago, is within the lower part of the **Transvaal Supergroup**, a thick sequence of mostly sedimentary rocks that were deposited within the Transvaal Sea on continental basement.

Landscape and landforms

The landscape reflects dissected terrain of the Crocodile River catchment, eroding back strongly along the Great Eastern Escarpment, and separating Mpumalanga's Lowveld from the Highveld plains. It is a dissected remnant of that which existed before the break-up of Gondwana. The once overlying Pretoria Group rocks have been removed and valleys have cut down through the Malmani dolomite and Black Reef quartzite, exposing the underlying Archaean granitic basement nearby. The residual massif of dolomite is crossed by north-trending linear structures, probably faults and dykes, which have allowed surface water to penetrate and erode along joints and fractures.

Rock formations

- The primary rock is a dark grey, finely bedded sediment commonly called 'dolomite' (a variety of limestone), consisting mainly of calcium-magnesium carbonate

($CaMg(CO_3)_2$). Another name for this is 'dolostone' although the term 'dolomite' is well entrenched in South African geology.

- At Sudwala the flat-lying dolostone layers vary from semi-massive to well layered and, in places, thin layers of whitish chert stand out.
- The dolostone was formed in shallow water from ultrafine sediment chemically deposited by blue-green algae (see below). This primitive marine life formed vast algal beds, now seen as rounded stromatolites of varying sizes.

Courtesy of Sudwala Caves

One of the many named cave formations, which are mostly upward-growing, pillar-like stalagmites; this one is called The Madonna. Grey, well-bedded dolomite (dolostone) with thin whitish chert layers is seen in the background.

Geological history

Around 2,500 million years ago, a miraculous (but natural) process started: using sunlight and by means of photosynthesis, blue-green algae (cyanobacteria) absorbed carbon dioxide dissolved in sea water and, as a waste product, released into the water a constant supply of unwanted oxygen. This simple process created a chemical imbalance in the calcium-rich sea water, resulting in extremely fine-grained calcium carbonate being chemically precipitated onto the algal surfaces, which then continued to grow. In this way, thick beds of carbonate rock where formed and, after taking on magnesium that was also in sea water, eventually became dolostone. Millions of years of burial resulted in the rock we see today, still with visible evidence of algal life.

Things to see and do

- **Do the cave tour:** The one-hour standard guided cave tour is suitable for young and old, and takes you about 600m into the mountain. You will see many amazing cave formations that are, in places, artistically illuminated and imaginatively named.
- **Do the Crystal Chamber tour:** Only for the adventurous, this takes place just once a month and lasts about 5 hours. It gets you wet and dirty, and takes you into a chamber filled with exquisite aragonite (another form of calcite) crystals about 2,000m from the entrance.
- **Visit the Dinosaur Park:** Right next to the caves, in a heavily wooded area, is a remarkable outdoor museum that has realistic life-size exhibits of prehistoric animals, always an attraction with the younger generation.

Realistic model of a Stegosaurus *dinosaur that lived in North America ~130 million years ago, seen in the educational Sudwala Dinosaur Park.*

28

Lone Creek Falls

PROVINCIAL HERITAGE SITE

One of several impressive waterfalls in the upper Sabie River Valley

The bottom of 68m-high Lone Creek Falls where the water plunges into an attractive pool that is backed by Malmani dolomite. This is a National Monument and now a Provincial Heritage Site. Inset: The full height of the falls is seen here, dropping off a minor escarpment capped by Bevets conglomerate.

The region around Sabie and Graskop is well known for its many waterfalls, a great tourist attraction in this spectacularly scenic part of Mpumalanga. One of the most popular is where the Lone Creek stream, a headwater tributary of the Sabie River, plunges over a 68m-high cliff into a pool below.

■ Location and getting there

- The town of Sabie lies at the meeting point of several routes. A little more than 60km northwest of Nelspruit, it's at the centre of the ever-popular Panorama tourist route. Lone Creek Falls lies about 9km west of Sabie, following the Old Lydenburg Road. Gravel roads that lead to Bridal Veil and Horseshoe Falls are signposted off this tarred road.

- GPS (pool at base of falls): 26°06′09″S; 30°42′34″E

Setting the scene

The small town of Sabie is situated in a deep valley close to the edge of the Great Eastern Escarpment. It is now the centre of a thriving timber industry based on vast plantations but it was the gold rush of 1872–1873 that first brought renown to the area. There are many waterfalls around Sabie and Graskop, and they can be grouped into two types based on their geological location (see p.188).

Regional geology

The area comprises the lower part of the almost flat-lying **Transvaal Supergroup**, from the basal **Black Reef Formation** to the top of the 2,000m-thick **Malmani Subgroup** (part of the **Chuniespoort Group**) and upwards into the overlying **Pretoria Group**. Above the **Malmani Subgroup**, forming the base of the **Pretoria Group**, is the **Rooihoogte Formation**, which comprises the basal **Bevets conglomerate**. This unit reflects an old erosional land surface that developed on top of the Malmani dolomite during a long period of ancient weathering, and marks the regional unconformity between the Chuniespoort and Pretoria Groups.

Landscape and landforms

The landscape around Sabie is part of the broader Great Eastern Escarpment. The Sabie River has cut deeply into the Black Reef-covered plateau west of the main escarpment edge, and above this, several tributaries have dissected the Malmani dolomite and the overlying Pretoria Group.

*The exposure behind the falls consists of two rock formations; from the base upwards there is a smooth face comprising giant domal stromatolitic Malmani dolomite (**A**); above this are overhanging and harder layers of Bevets conglomerate (**B**).*

An enlarged portion of the so-called 'bread and butter' dolomite, part of a giant domal stromatolite (location indicated on top photo). It is composed of alternating layers of light-coloured dolomite and thinner, harder chert.

This specimen of very hard Bevets conglomerate is from Horseshoe Falls. Detailed examination shows that it is made up of fragments of older chert recemented by silica to form a concrete-like mass that resists erosion.

Two groups of waterfalls can be recognised, where strong headward erosion has reached major knickpoints. The first group comprises waterfalls produced by resistant Black Reef quartzite, where streams drop off the escarpment edge, examples being Sabie, Mac-Mac and Panorama Falls at Graskop. The second group is higher up in the geological sequence, where a hard ledge of Bevets conglomerate has produced Lone Creek, Bridal Veil and Horseshoe Falls.

Rock formations

- Malmani dolomite: The light brown banded rock seen at the lower part of the Lone Creek waterfall cliff consists of pale dolomite layers alternating with thin whitish chert bands. The rock's distinctive appearance has led to its informal name, the 'bread and butter dolomite', and it provides a good regional marker horizon. Here the very characteristic giant domal stromatolitic structures are at the top of the Malmani dolomite sequence (see Sudwala Caves, p.181–184).

- Bevets conglomerate: The upper part of the waterfall cliff forms a resistant overhang that is made of a chert fragment breccia recemented by secondary silica. This rock formed by the large-scale accumulation of chert fragments weathered out of the dolomite during a long period of erosion, before deposition of the Pretoria Group.

Small caves have formed in blocky, jointed dolomite alongside the road to Lone Creek Falls.

The Mac-Mac goldfield and Pilgrim's Rest

In 1871 prospectors Tom McLachlan, George Parsons and others were working in the mountainous region around what is now Graskop and Sabie. By 1872, encouraging alluvial finds had been made and diggers flooded into the area, heralding the start of the first major gold rush in South Africa. By mid-1873 the diggings were proclaimed and more than 1,000 mostly foreign diggers were active in what became known as the Mac-Mac goldfield. In September 1873, riches were struck in the Pilgrim's Rest goldfield by Alec 'Wheelbarrow' Paterson, forever changing this part of Mpumalanga, and leading to the start of the much larger Transvaal Drakensberg goldfield.

Gold is mainly associated with numerous horizontal, sulphide-rich, hydrothermal 'flat-reefs' (parallel to the bedding) and vertical 'cross-reefs' (cutting across the bedding) within the Malmani dolomite, and soon underground mining took over, with Pilgrim's Rest becoming the centre of gold production. Large-scale mining at Pilgrim's Rest finally stopped after more than 100 tons of gold had been recovered; the town was declared a National Monument, and has been preserved as a 'living museum' commemorating the local gold-mining industry. Close to Sabie, Glynn's Lydenburg Mine eventually became the main hardrock gold producer, yielding more than 38 tons of gold over a 55-year working life. A lot of gold remains in the ground but it will require a much higher gold price to make extraction affordable.

Things to see and do

- **Walk to Lone Creek Falls:** The waterfall was made a National Monument in 1983 and is now a Provincial Heritage Site. It is easily accessed and has tourist facilities. It is only about a 150m walk from the car park to the base of the falls, through indigenous forest.
- **See Horseshoe Falls and Bridal Veil Falls:** These waterfalls are of identical geological origin: both streams fall off the silicified Bevets conglomerate, overlying bedded Malmani dolomite. Horsehoe Falls is so named as the stream splits into two; the 70m-high Bridal Veil Falls (which includes a 750m uphill walk) is named for the veil-like curtain created by the falling water. Climbing around the falls is unsafe and not recommended.

Horseshoe Falls, another Provincial Heritage Site, is so named as it splits into two, in horseshoe fashion. Here hard silicified Bevets conglomerate is the main rock formation and lies stratigraphically above the Malmani dolomite. This is the same rock that forms the upper resistant ledges at the other falls in the area.

- **See Sabie Falls and gorge:** When in Sabie it's worth having a look at the local falls to see the 'escarpment edge' style of waterfall. Here the now larger Sabie River plunges 73m over a very resistant ledge of flat-lying Black Reef quartzite into the spectacular gorge below, on its way to the Eastern Lowveld.
- **Drive Long Tom Pass:** No visit to Sabie would be complete without a drive up this scenic road on the way to Lydenburg, and a stop at the historic Long Tom monument and replica cannon. This was the route used by the early wagons coming from Delagoa Bay (now Maputo) into the interior.

29

God's Window

KRUGER TO CANYON BIOSPHERE RESERVE

A must-see stop in the land of waterfalls on Mpumalanga's famed Panorama Route

Looking south from the 'gorge' viewsite at God's Window. The vertical whitish formation on the right is horizontal upper Wolkberg Group quartzite that overlies thickly vegetated granite-gneiss of the Eastern Lowveld, on the left. In the distance, probable Black Reef Formation quartzite forms the plateau top.

This almost unrivalled viewsite reaches 1,730m, the total drop to the Eastern Lowveld far below is around 800m, and on a clear day you can see 100km across the Kruger National Park to the distant Lebombo Mountains bordering Mozambique. Here you can see how the underlying geology and land-forming processes have combined – and serendipitously now provide us with spectacular views.

■ Location and getting there

- God's Window is located on a circular drive about 9km north of Graskop. To get there take the R532 to Motlatse River (formerly Blyde River) and, after about 2km, turn right onto the circular R534. The road is tarred, well signposted and rejoins the R532 further north. From the car park it is a short walk to the 'gorge' viewsite, then there's an uphill footpath to the Lowveld viewsite. About 1.5km further north along the main road there is the more accessible Wonder View viewsite. The drive will take you to several other geological attractions in the area.

- GPS (God's Window 'gorge' viewpoint): 24°52′36″S; 30°53′20″E

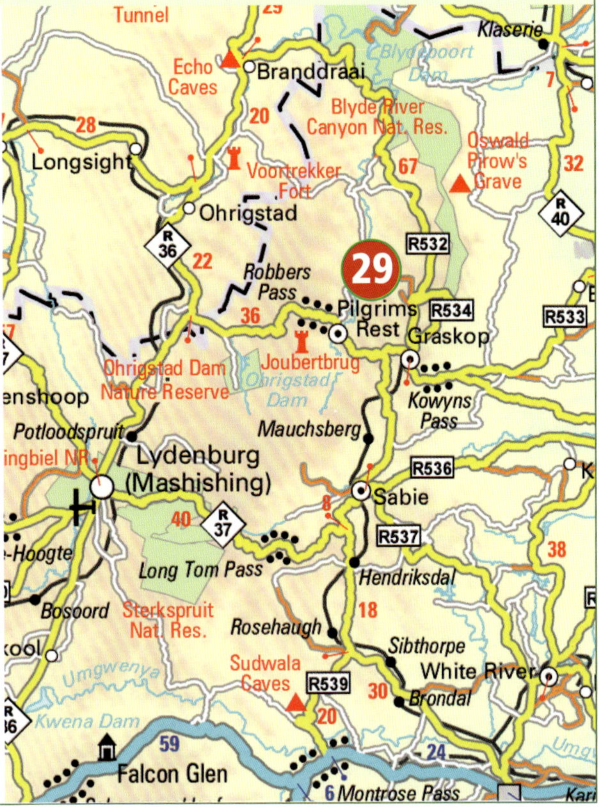

Setting the scene

There are actually two viewsites, one down a narrow gorge that gives you a view southwards along the precipitous edge of the escarpment (as pictured opposite), and another that looks eastwards across the Lowveld from a higher vantage point. The latter takes you through a remnant of indigenous fern and aloe forest leading to the magnificent view. At an elevation of around 1,730m the outlook, particularly in summer, is often misty or hazy. This area is managed as part of the Motlatse Canyon Provincial Nature Reserve.

Part of Driekop Gorge between God's Window and Graskop. The top grassy layer across the gorge is mapped as Black Reef Formation quartzite which overlies more massive upper Wolkberg Group quartzite.

Panorama Falls and gorge lie immediately south of Graskop village. Horizontal Black Reef Formation quartzite overlies very similar upper Wolkberg Group quartzite, and they have been eroded into a deep gorge on the edge of the escarpment.

Regional geology

This is part of the Great Eastern Escarpment and its geological setting is the easternmost part of the large Transvaal Basin. **Black Reef Formation** quartzite overlies the top of the **Wolkberg Group**, a precursor to the **Transvaal Supergroup** (see Mpumalanga Overview, pp.162–163). The Wolkberg Group, probably around 2,700 million years old, filled an early basin of limited extent and is better seen in the vicinity of the Blyde River Canyon (see pp.199–203).

Landscape and landforms

The Great Eastern Escarpment is one of the world's most impressive landscapes, and God's Window is one of its highest points (1,730m). In the Mpumalanga region the escarpment extends more or less continuously from well south of Kaapsehoop to northwest of Blyde River Canyon, and is shown on maps as the Drakensberg. Further to the northwest the mountains are known as the Wolkberg.

The layered formations that exist along this escarpment generally dip very slightly to the west, and they overlie the more easily eroded granites and gneisses of the Basement, seen at the base of the escarpment and on the Lowveld plains. The escarpment that we see today is only a part of the major escarpment that occurs all around southern Africa. The scarp has been aggressively eroded back from the early coastline, and east- and north-

A close-up view of almost flat-lying Black Reef Formation quartzite at the viewsite on the edge of Panorama Gorge, the same formation as that seen at Kaapsehoop.

Lisbon and Berlin Falls

These falls differ from others in the area in that they are located on westward-flowing streams and do not cascade off the Great Escarpment. About 3km off the R532, on the Lisbon River (which is a tributary of the Motlatse River), the multiple streams of Lisbon Falls cascade over a 92m cliff. Berlin Falls (a National Monument and now a Provincial Heritage Site) – also known as Berlyn Falls – is on the nearby Waterval Spruit and the 80m drop off a scarp of hard quartzite is impressive. The dip of the quartzite strata is shallow and they are probably part of the upper Wolkberg Group rather than the Black Reef Formation. It is thought that both these waterfalls have formed along a line of north-trending, fault-related scarps.

Not far from God's Window and the escarpment edge is the multiple, 92m-high Lisbon Falls, dropping off Wolkberg quartzite and draining westwards.

A walk to the Lowveld viewsite at God's Window takes you through a remnant patch of rocky montane rain forest, growing on lichen-covered Wolkberg quartzite.

Berlin (or Berlyn) Falls are found on the westward-draining Waterval Spruit, where they fall some 80m from a cliff of Wolkberg quartzite.

flowing drainages cutting back into it are still very active. The geological map reveals evidence of a north-northeast-trending, dyke-filled structure close to the escarpment edge, and God's Window 'gorge' is likely to coincide with this structure (see p.190).

Rock formations

● The underlying, almost horizontal and well-bedded quartzite is geologically mapped as being the upper Wolkberg Group. One could be forgiven for thinking that it is part of the more widespread and overlying thin Black Reef Formation that occurs regionally across the plateau. There is considerable similarity in appearance between the upper Wolkberg rocks and the Black Reef quartzite, as seen at Bourke's Luck Potholes. Both formations were formed from clean, reworked quartz sand under similar shallow-water conditions, and only very detailed geological mapping can make a distinction between them.

Things to see and do

- **Walk to the viewsites:** A short stroll to the gorge viewsite is a must-do. This gives an excellent view of the escarpment to the south and a look at the bedded Wolkberg quartzite rock formation. A walk to the Lowveld viewsite through indigenous forest is well worth the effort.
- **See The Pinnacle and Driekop Gorge:** Located on the edge of another gorge north of Graskop, this 30m-high column of blocky, jointed Wolkberg quartzite has been isolated by erosion from the escarpment edge, and close by you will see collapsed blocks of quartzite – ages ago this pillar was part of the plateau.
- **Graskop Gorge and Panorama Falls:** These are immediately south of Graskop, off the road to Kowyns Pass and Hazyview (R533), where the gorge is the site of adventure activities. The wall of the gorge displays an excellent vertical section of upper Wolkberg and Black Reef quartzites. The falls are some 80m high.

This is 'The Pinnacle', a well-known squarish (strongly jointed) pillar of Wolkberg quartzite that has been left isolated by erosion in Driekop Gorge.

30 Bourke's Luck Potholes

KRUGER TO CANYONS BIOSPHERE RESERVE

A wonderful demonstration of the erosive action of small turbulent rivers

The best display of deep, water-worn potholes is seen at the confluence of the Motlatse (formerly Blyde) and Treur Rivers, the latter seen here in the background. Pedestrian bridges and pathways over and around the site reveal the scale of this geological wonder.

This is one of South Africa's most remarkable geological phenomena where, over millions of years, swirling water in the fast-flowing Motlatse and Treur Rivers has scoured out deep cylindrical cavities in the hard bedrock. The place is named after prospector Thomas Bourke who reportedly worked here in the 1880s and allegedly made a lucky gold strike. Today, the village of Moremela is rapidly expanding alongside the iconic geosite.

Location and getting there

- The potholes are located about 35km north of Graskop, close to the R532 and at the southern end of the famous Blyde River Canyon Nature Reserve, now renamed the Motlatse Canyon Provincial Nature Reserve. Their position at the confluence of the Treur River ('river of sorrow') and Blyde River ('river of joy') marks the start of the renowned river canyon. There is a small charge to enter the reserve.
- GPS (first bridge across gorge): 24°40'27"S; 30°48'38"E

Setting the scene

In about 1880 Thomas Bourke, a gold prospector who had moved away from Pilgrim's Rest, reportedly struck it lucky when prospecting along this stretch of the Treur River. Legend has it that he found rich alluvial gold in the potholes, and, although some stories suggest he found very little gold, the farm got the name of Bourke's Luck. In any event, gold was later found in sulphide-rich hydrothermal veins within fractures in the quartzite, close to a north-trending dyke. Nearby Bourke's Luck gold mine closed down only in 1955, having produced 4.5 tons of gold from a number of spread-out workings, plus by-product copper.

From the first bridge, looking upstream (southwards) along the remarkably narrow Motlatse River. The straightness of the river suggests that its course is the result of a linear geological structure.

Regional geology

The setting here – as with the other geological sites in this part of Mpumalanga – is part of the Great Eastern Escarpment, showing the eastern

At the potholes the sequence is fairly simple, consisting only of stacked quartzite layers: probably Black Reef (on top) and Wolkberg (at bottom).

The flagstone paving (quarried nearby) shows remarkable angular casts of halite crystals that indicate an original salt-rich shale deposit within the sedimentary sequence.

The Treur River as it cascades into the Motlatse River before reaching the large potholes. There is an old mine working in the distant background.

part of the large Transvaal Basin. Here, the underlying formation is the **Black Reef Formation**, which does not exceed a thickness of a few metres. It forms the basal unit of the **Transvaal Supergroup** and is around 2,650 million years old (see Mpumalanga Overview, pp.162–163). The Black Reef itself lies upon much thicker quartzites of the upper part of the

Along the Treur River there are eroded pinnacles of Black Reef quartzite, formed by jointing and erosion, and possibly the starting point for pothole formation.

slightly older **Wolkberg Group** (see Blyde River Canyon, pp.199–203), which formed under similar depositional conditions. A notable aspect of the geology, as can be seen on Google Earth, is the strong north-northeast structural trend, including the Dientje dyke, which has influenced the alignment of gold deposits.

Landscape and landforms

There is good evidence to indicate that here the north-northeast direction of the Motlatse River is controlled by a major structure – a probable fault zone in the flat-lying rock strata. A landscape of weirdly eroded pillars of quartzite is not uncommon in the area, created by preferential rock weathering and different rates of erosion, and this is certainly influenced by the pattern of rock jointing and assisted by the higher rainfall along the escarpment edge. Similar conditions formed the peculiar rock formations at the Kaapsehoop Escarpment further south (see pp.171–175). The process of erosion has gone one step further to create the potholes at Bourke's Luck. Where the Motlatse and Treur Rivers meet on the shallowly dissected landscape, there was further cutting down into the quartzite and the creation of bowl-like hollows that entrap river boulders and pebbles (see text box on p.198). It is speculated that this vigorous erosional process would not have been conducive to the accumulation of fine-grained gold.

A view of the Motlatse River from the second bridge, close to the biggest potholes.

Rock formations

- What you see are magnificent exposures of eroded quartzite, either Black Reef or upper Wolkberg, ground smooth by countless thousands of years of river erosion.
- The reddish and blackish coloration is due to staining by iron and manganese oxides, respectively.
- Originally the quartzites formed as very extensive blanket-like layers of mainly clean quartz sand and, in places, of mud or grit and pebbles, along a shallow coastline (including evaporative mud flats) at a time when the Transvaal Basin was starting to fill.

How do potholes form?

Under precisely the right conditions, potholes can form in rivers when the debris – small boulders, pebbles and sand – being transported by the current wears away the rocky river floor in a process of abrasion. Loose material becomes trapped, albeit temporarily, in bedrock hollows and is rapidly swirled around by eddies in the fast-flowing water, especially where the river has been funnelled into a narrow channel. The hollows gradually grow wider and deeper, and the loose rocks can drill down several metres to produce large cylindrical potholes, also called 'cauldrons'. In many places, adjacent potholes eventually overlap and merge. While the fine debris is continually being carried away by the strong stream of water, new pebbles and sand are regularly being added. For this to take place a delicate equilibrium needs to be maintained between the flow of the river, the size of pebbles and the nature of the bedrock. Bourke's Luck potholes have it all.

Things to see and do

- **Walk around the pathways:** There are several pedestrian paths and bridges to follow and a host of intriguing primary sedimentary (depositional) and later erosional rock features to see. Regrettably, there was (at the time of visiting) no geological information available at the site.
- **Look out for crystal casts:** On the locally derived paving stones look out for very distinctive, squarish casts of salt crystals (halite) that could only have formed by evaporation of strongly saline sea water on mud flats. You will also see well-preserved ripple marks and possibly ancient mud cracks in the pavement.

Downstream from the potholes marks the start of the Blyde River Canyon, which rapidly deepens northwards; seen here from the second bridge.

31 Blyde River Canyon

PROVINCIAL NATURE RESERVE

One of the largest canyons on Earth, a magnificent demonstration of deep river incision

Walter Knirr / Images of Africa

Looking northwards towards the Blyde River Dam wall, with the start of the Wolkberg range in the far distance. Except for a thin layer of flat-lying Black Reef quartzite on top, all the formations are part of the thick, proto-basinal sequence of the Wolkberg Group.

This superb canyon is one of the finest spectacles in southern Africa and probably one of the country's most photographed scenes. Now part of the newly named Motlatse Canyon Provincial Nature Reserve, it is said to be the third-largest canyon on Earth and the world's longest 'green' canyon, thanks to its lush vegetation.

Location and getting there

- There are several spectacular viewpoints, from both the top and bottom of the canyon. For a top-down view take the R532, starting about 50km north of Graskop. For a bottom-up perspective you need to be on the Lowveld side of the Great Escarpment and make your way to the Swadini Resort within the Motlatse Canyon Provincial Nature Reserve, about 11km off the R531. A further 3.5km takes you to the wall of the Blyde River Dam.
- GPS (dam wall): 24°32′11″S; 30°47′54″E

A different view of the Wolkberg from near the Blyde River Dam, which was created by damming the deeply incised Motlatse River. The formations are all part of the alternating sequence of quartzites and shales of the Wolkberg Group.

Setting the scene

The canyon starts immediately after Bourke's Luck Potholes (see pp.195–198) and ends nearly 30km downstream near the Blyde River Dam wall. The Blyde River (now renamed the Motlatse River) is a tributary of the Olifants River; the approximately 300km^2 surrounding nature reserve, proclaimed in 1965, is now called the Motlatse Canyon Provincial Nature Reserve. Here the geology of the Great Eastern Escarpment is superbly revealed in both the deeply incised gorge of the Motlatse River and in the surrounding mountains, where it has produced wonderful panoramas.

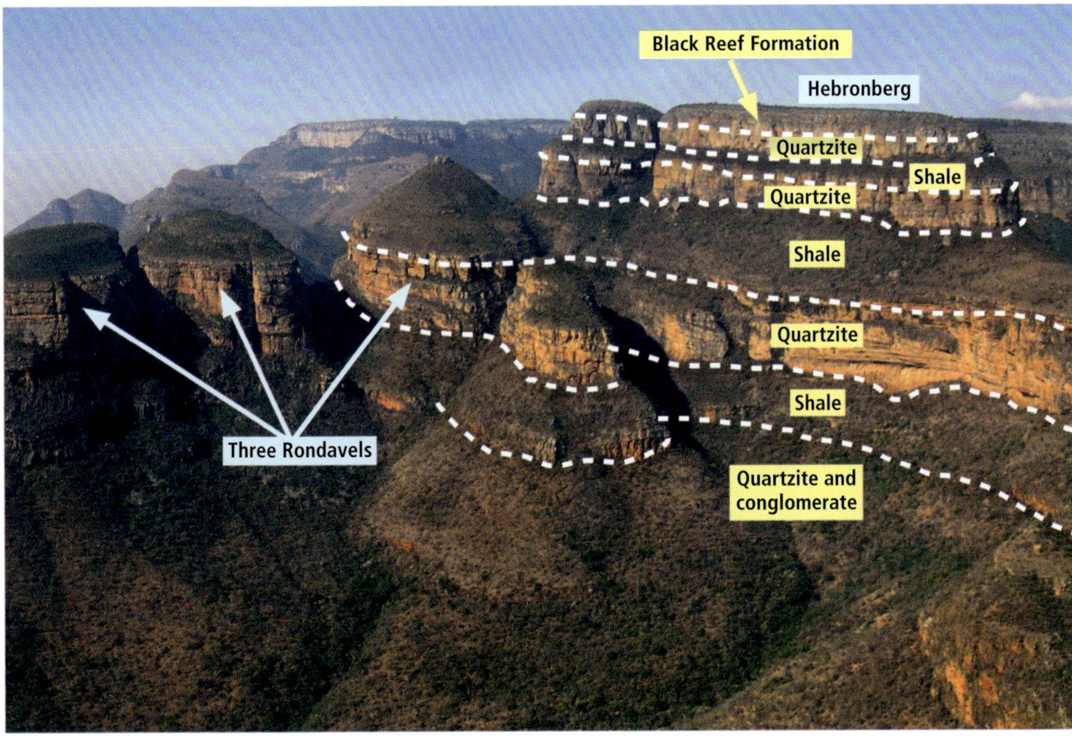

One of the classic views at the Blyde River Canyon is the iconic 'Three Rondavels', remnants of once covering formations, their conical 'roofs' of more easily worn down shale, and their 'walls' of more resistant quartzite. The geology is illustrated in the natural geological cross-section provided by the neighbouring massif of Hebronberg, capped by a layer of Black Reef Formation quartzite. Alternating quartzite and shale formations of the Wolkberg Group, here about 600m thick, are indicated by dashed white lines. The lowermost Wolkberg formations are not visible.

Regional geology

This is the easternmost part of the Transvaal Basin, a depository that started to form around 2,700 million years ago (see Mpumalanga Overview). At the top of the dissected plateau the formation immediately underlying the surface is mostly mapped as the 2,650-million-year-old **Black Reef Formation**, deposited blanket-like on the sedimentary formations stratigraphically underlying and preceding the **Transvaal Supergroup**. The older, underlying strata are of the precursor basin of the **Wolkberg Group**, comprising mostly alternating layers of quartzite and shale, here at their thickest and magnificently exposed in the walls of the canyon. All formations dip very shallowly to the west, as can be seen on the flat-topped mountains across the canyon.

Landscape and landforms

The upper escarpment plateau is a remnant of the ancient erosion surface that once existed above the African land surface. It has been deeply incised by the north-flowing Motlatse River and its tributaries, and follows a sinuous and ever-deepening northwards course for more than 30km before entering the Eastern Lowveld plains. The plateau has been cut down to a depth of 600–800m, resulting in numerous brightly coloured cliffs and scarps that provide the most spectacular sights. Along the escarpment edge the partly dissected surrounding mountains reach elevations of 1,944m at Mariepskop, 1,794m at Mogologolo and 1,767m at Hebronberg, while the elevation of the

The weathering role of lichen

In many places in South Africa, particularly where it is often damp, weathered rock outcrops sport irregular, almost paint-like white, grey or orange patches of lichen, which may completely cover the rock. Lichens are widespread, long-lived and incredibly tough, occurring in some of the most extreme environments on Earth, and are an important agent in the weathering of rocks. They are a composite, two-part organism comprising a fungal species and a photosynthesising algal species bound up intimately in a symbiotic relationship, providing each other with nutrients, moisture and protection. As primary colonisers of bare areas, their action leads to weathering of the underlying rock surface by both physical and chemical processes: minute mechanical disruptions within the rock mass (penetration, expansion and contraction) and the excretion of organic acids (particularly oxalic acid) that effectively dissolves and corrodes minerals. Lichens have different growth forms, ranging from brightly coloured crusty coverings and papery, leaf-like outcrops to bushy, beard-like growths.

Vivid orange, white and grey lichens have colonised these quartzite outcrops. The pale grey leafy lichen (to the right), in particular, is flourishing. Lichen is a slow but potent organic agent in the weathering of rock outcrops.

river beyond the dam wall is about 600m. The best-known view is of the iconic massifs known as the Three Rondavels.

The immense canyon is the erosion product of the relatively small Motlatse River, which originates near Pilgrim's Rest, and which has notably maintained its earlier sinuosity. Clearly, uplift of the terrain has been matched by the strong down-cutting action of the river. Well developed north-south and southeast-northwest structures have also influenced erosion, such as at the Three Rondavels. This iconic landscape is an excellent demonstration of differential erosion of softer shale and harder quartzite formations.

Rock formations

- Only the well-bedded and almost flat-lying Black Reef Formation quartzite on the upper plateau is visible close up. Geological interpretation suggests that this may be the Langkrans Formation quartzite of the upper Wolkberg, which is of identical appearance.
- The Wolkberg Group has been subdivided into eight formations of mostly alternating quartzites and shales, as seen in a view across the canyon from the escarpment edge.
- These alternating rock formations represent changing environments of deposition under river and shallow marine-to-deltaic conditions within an unstable early sedimentary basin.

Things to see and do

- **See the Three Rondavels:** This is the best view of the canyon from the top edge, where you can see most of the Wolkberg Group, layer by layer, and the three semi-isolated peaks across the canyon, and get spectacular views of the mountains and the Blyde River Dam.
- **Take in the Lowveld view:** Another spectacular viewsite across the canyon is about 5km further south, also off the R352. From here you can catch a glimpse of the Lowveld between the highest peaks.
- **Visit the dam wall:** A 71m-high dam wall holds back the river to form a large dam that was opened in 1975 at the confluence of the Ohrigstad and Motlatse Rivers. A look-out point provides exceptional views of the gorge's spectacular rock formations, and boat trips can be arranged.
- **Drive up Mariepskop:** From the Acornhoek side of the escarpment (off the R40) you can drive up a tarred road to the top of this impressive mountain and have a superb view of the Eastern Lowveld.

A view looking eastwards across the deeply incised canyon towards Hebronberg (1,767m); the strongly jointed quartzites (on the right) have certainly assisted erosion.

Looking across the canyon towards the 1,944m-high landmark of Mariepskop, and the remarkably small Motlatse River at the bottom of the valley, part of the Motlatse Canyon Provincial Nature Reserve. Over time an enormous amount of erosion has taken place. You can get to the top of Mariepskop from Acornhoek, off the R40.

LIMPOPO

OVERVIEW OF LANDSCAPES AND GEOLOGY

Limpopo is a region of considerable geological antiquity and variety, and of diverse landscapes. These include the Limpopo Lowveld in the north, the elongated massif of the Soutpansberg, the Waterberg Plateau in the west, and the northern part of the Great Escarpment. The focus in this chapter is on Geosites 32–38, shown on the national maps (pp.12–13), and on the regional landscape and geological maps (pp.206–207).

The major geological formations of the region are listed below:

Limpopo Complex
(Archaean)

North of the Soutpansberg is a region of highly complex geology, comprising granites, gneisses, migmatites and a variety of deformed, strongly metamorphosed rocks, all known collectively as the **Limpopo Metamorphic Complex**. This challenging geological terrain, also called a 'mobile belt', started to form very early in Earth's history, around 3,400 million years ago, when the young Kaapvaal and Zimbabwe Cratons began moving towards each other, creating a wide zone of intense folding in between. This ancient collision of minor continents is viewed as one of the world's oldest mountain-building events and it had far-reaching consequences for the geology of southern Africa. Near Musina is the spectacular **Sand River Gneiss** exposure (**Geosite 34**). Topographically, this region forms the low-lying Limpopo Lowveld, a broad, mostly granitic or gneissic plain drained by the Limpopo River and its sandy tributaries.

Pietersburg Group
(Middle Archaean)

This comprises the **Pietersburg Greenstone Belt**, a northeast-trending sequence of ultramafic, mafic and sedimentary formations located southeast and east of Polokwane. The narrow belt stands out in positive relief from surrounding **Basement** granitic gneisses, and has the distinction of hosting South Africa's first gold mine at Eersteling (**Geosite 36**). The sequence has been metamorphosed to various schists and amphibolites, and there are also iron formations, ferruginous cherts and some felsic volcanics. The hardrock gold mineralisation is typically associated with steep shear zones and quartz-carbonate-sulphide alteration. At around 3,000 million years in age, the rocks are somewhat younger than the **Barberton Greenstone Belt**.

Basement Granite-Gneiss
(Archaean)

There are no geological sites featured in this broad geological terrain but it needs a mention. It comprises a variety of granites and gneisses, including numerous metamorphosed remnants of greenstone. The intensity of regional metamorphism increases northwards towards the **Limpopo Complex**. Geomorphically, the granites and gneisses form the broad Polokwane Plain, which on its eastern side becomes part of the Great Escarpment, below which are similar granites and gneisses of the Eastern Lowveld.

Transvaal Supergroup
(Late Archaean to Early Proterozoic)

The lower part of this thick sequence, in particular the **Black Reef Formation** and **Malmani Subgroup**, is outlined in the Mpumalanga Overview (pp.162–163). The prominent Black Reef quartzite continues to provide the now tilted base for the same Malmani dolomite formations. Similar geology continues from the **Wolkberg Group** in the southeast, creating the northwest-trending Strydpoort Mountains, after which the formations turn abruptly northwards east of Mokopane to form the rim of the Transvaal Basin. The high ground is a remnant of the ancient land surface that predated the break-up of Gondwana, now dissected and lowered. The many famous caves in Makapans Valley (**Geosite 37**) are, however, very much younger.

Bushveld Complex
(Early Proterozoic)

This enormous magmatic complex makes an appearance in the south as part of the Eastern Limb (see p.210), and west of Mokopane as the Northern Limb, but no geosites are featured here from the latter. Geomorphically the Eastern Limb forms the northern part of the lower-elevation Bushveld Basin (see Dwars River Chromite, **Geosite 32**; shown on Mpumalanga regional maps).

Waterberg Group
(Early Proterozoic)

To the southwest lies the elevated Waterberg (or Palala) Plateau, comprising a 5,000m-thick sequence of flat-lying, well-bedded coarse sandstones and minor conglomerates (the same colour as the **Soutpansberg Group** on **Figure 26**). In the south and east they are fronted by the impressive step-like Waterberg Escarpment. These rocks were deposited around 2,000 million years ago on a stable braided alluvial flood plain, the sediment being derived from the ancient Limpopo highlands in the northeast shortly after the intrusion of the **Bushveld Complex**. They are called 'red beds' because of their characteristic mottled, reddish-brown to maroon coloration due to pervasive rusty iron oxides, and they are probably the oldest of their kind on Earth. The presence of iron oxide in sandstone provides evidence of free oxygen being available in the atmosphere for the first time, which steadily increased and ultimately enabled life to evolve on land. Excellent examples are seen at Waterberg's Kloof Pass (**Geosite 38**).

Soutpansberg Group
(Early Proterozoic)

The ancient Soutpansberg range forms an imposing elongated massif directly north of Makhado and rises to over 1,700m, being much more resistant to erosion than the older bordering granitic gneisses. It is made up of a 7,000m-thick sequence of upper sedimentary and lower basaltic volcanic formations, estimated to have been formed around 1,900 million years ago. These rocks were deposited within a faulted graben structure, now revealed by many east-west-trending strike-faults, but its origin – possibly a failed rift valley – is still being debated. Dominated by the 1,500m-thick, hard and pebbly **Wyllie's Poort Formation**, these formations are tilted to the north and have been repeated by step-faulting, resulting in a landscape of steep scarp-slopes and gentle dip-slopes, as seen around Wyllie's Poort (**Geosite 33**). As in the Waterberg, the original reddish to maroon, iron-rich sandstone (now quartzite) is also an example of oxygenated red beds (see **Waterberg Group**).

Karoo Supergroup
(Carboniferous to Jurassic)

The Tuli Basin forms an east-west-trending trough in the vicinity of the Limpopo River, and is an outlying Karoo-age basin filled with a restricted sequence of sedimentary rocks. In Zimbabwe and Botswana it is largely covered by basaltic lavas related to the final period of Karoo igneous activity. However, in the Limpopo Valley the uppermost sedimentary rocks are well exposed: the windblown **Tshipise Member** (the time equivalent of the **Clarens Formation**) and the **Red Rocks Member** (which correlates with the alluvial flood plain deposits of the **Elliot Formation**) are separated from each other by a notable ancient weathering surface that formed around 190 million years ago. At Mapungubwe (**Geosite 35**) a unique and historic landscape is revealed of sandstone mesas, isolated buttes and minor scarps, separated by wide sandy valleys. It is a Cultural World Heritage Site.

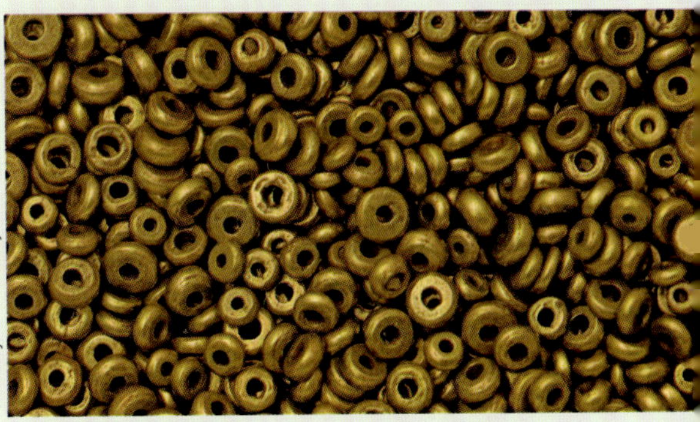

Courtesy of University of Pretoria (UP)

Countless small handmade gold beads found at the Mapungubwe World Heritage Site, and used for necklaces.

Figure 25: Landscape map of the Limpopo region. This region is made up mainly of the Polokwane Plain in the centre (yellow and green) and the lower-lying Limpopo Lowveld in the north (blueish-grey). Between these are the massifs of the Soutpansberg and Blouberg (yellow and brown). The deeply dissected Great Escarpment is in the east, the edge of the Waterberg Plateau in the west, and part of the Bushveld Basin in the south.

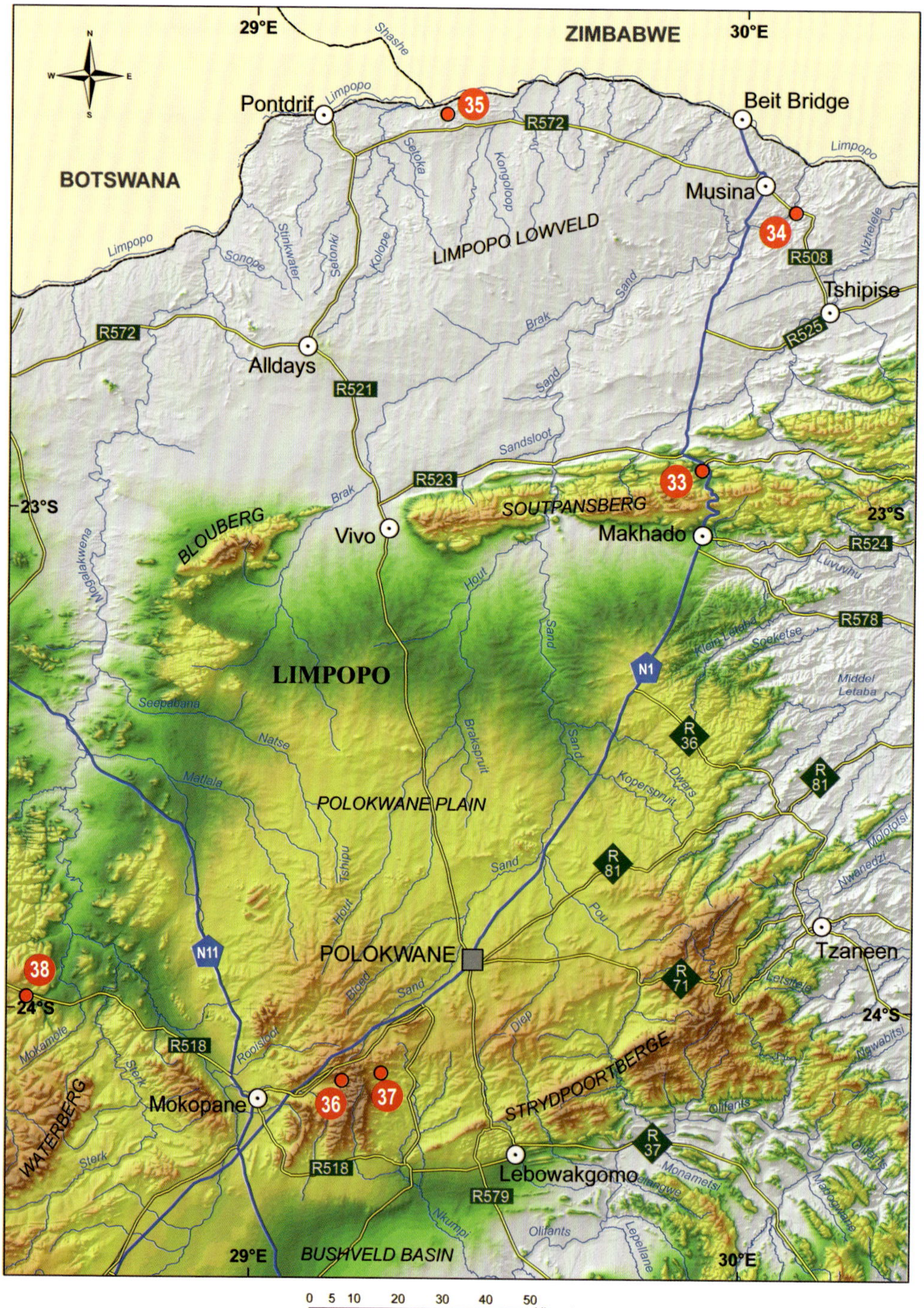

Figure 26: Geological map of the Limpopo region. This region highlights the close relationship between geology and landscape. In the north the rocks of the Limpopo Metamorphic Complex (shown in Greenstone Belts colour) and the Beit Bridge Complex form the lower ground. The Soutpansberg and Waterberg Groups are very similar in age and character, and both form mountainous landscapes, while the Polokwane Plain is mostly underlain by granite-gneiss of the Basement.

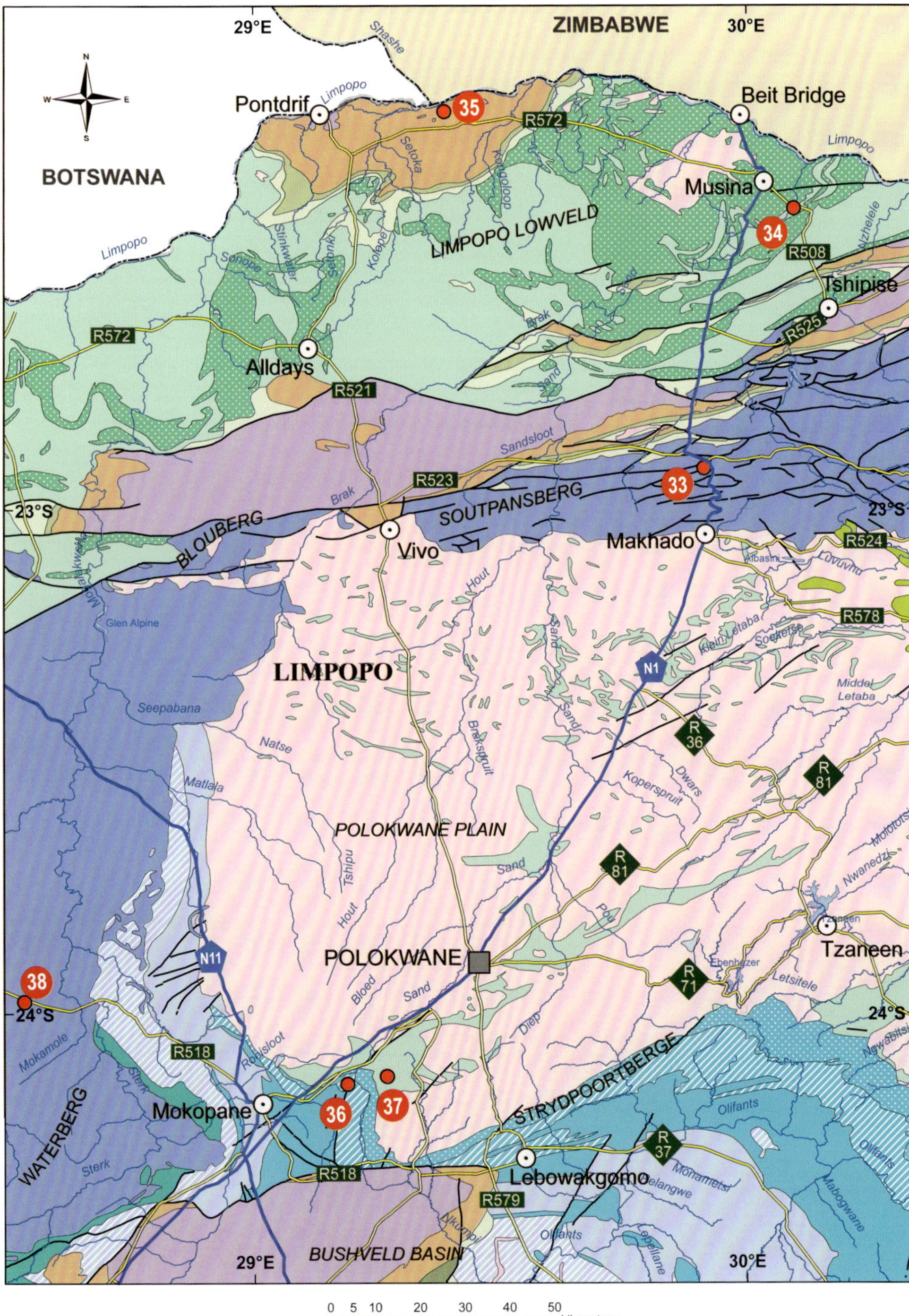

• A stratigraphic key to this simplified regional geological map appears on the inside covers

32 Dwars River Chromite

PROVINCIAL HERITAGE SITE

A world-famous exposure in the Bushveld Complex
showing an enigmatic igneous relationship

The famous Dwars River Chromite occurrence is one of South Africa's few geological National Monuments. Here the small Dwars River has formed a narrow defile within a zone of vertical fractures, and in turn this has created magnificent exposures of alternating igneous rock layers. The hills in the background contain an outlier of the platinum-rich Merensky Reef that is higher up in the igneous rock sequence.

Although the Dwars River Chromite geosite is in Limpopo province, for convenience its location is shown on the Mpumalanga maps (pp.164–165); and here the Bushveld Complex is best seen in its Eastern Limb, a must-see place on any geologist's wish list. There are many fascinating exposures, one of the most intriguing being the unique chromite occurrence at Dwars River, world-renowned for the intimate association between layers of black chromite rock and whitish anorthosite, an igneous rock that consists almost entirely of plagioclase feldspar.

Location and getting there

- This site is situated in the scenic Steelpoort Valley area between Roossenekal and Steelpoort, near Kennedy's Vale, and about 9km off the R555. You get there by taking the R577 to Lydenburg and must turn off the tarred road (not signposted) at the Dwars River bridge, from where you can see the site, located on the property of Two Rivers Platinum Mine.
- GPS (at plaque monument): 24°54'39"S; 30°06'13"E

Setting the scene

The Bushveld Complex comprises two major but very different related magmatic intrusions: the mafic Rustenburg Layered Suite (RLS) and the overlying felsic Lebowa Granite Suite. The RLS in the Eastern Limb provides exceptional exposures thanks to the strata having been uplifted and gently tilted westwards, revealing various types of igneous rock layers. These were intruded as magma into the crust over vast distances, are geologically distinctive, and form a 9km-thick systematic rock sequence (see stratigraphic column on p.210). While much is known about the RLS, the Dwars River Chromite occurrence, first brought to world attention by American geologist Edward Sampson in 1932, is famous because of the extremely rare association of two very different rock types: the igneous interlayering of dark chromite layers within light anorthosite layers. Which rock was formed first, and how did one intrude the other?

A close-up look at some of the black chromite-rich layers within usually thicker, greenish-white anorthosite layers.

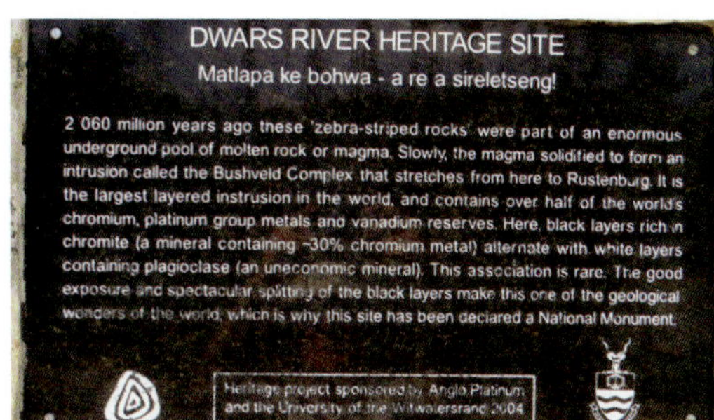

An informative engraved plaque has been erected overlooking the chromite exposure, and now a Provincial Heritage Site, sponsored by platinum mining company AngloPlatinum and the University of the Witwatersrand.

Figure 27: The Eastern Limb of the Bushveld Complex.

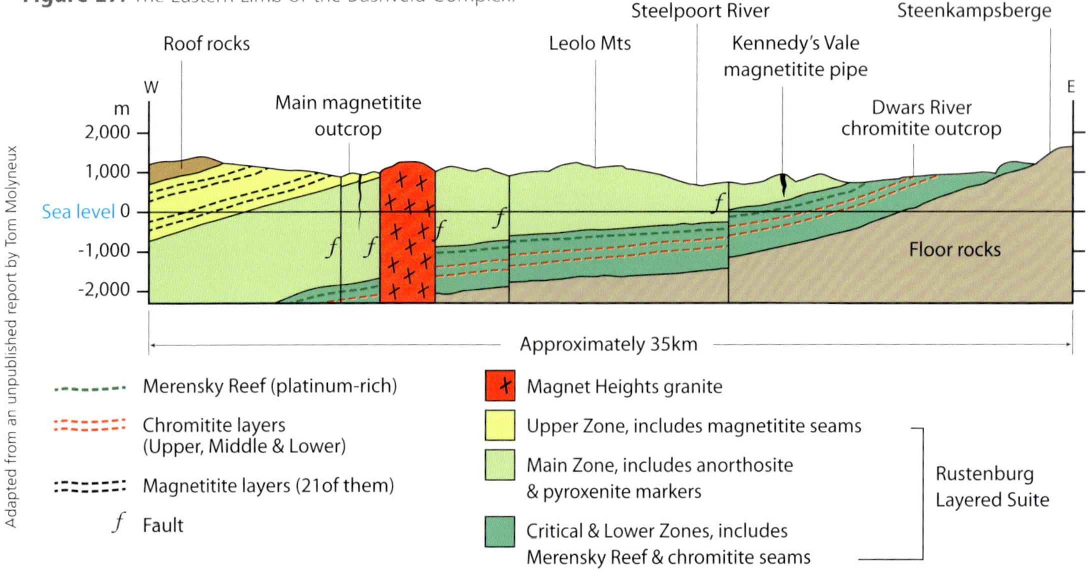

Figure 28: The Eastern Limb of the Rustenburg Layered Suite.

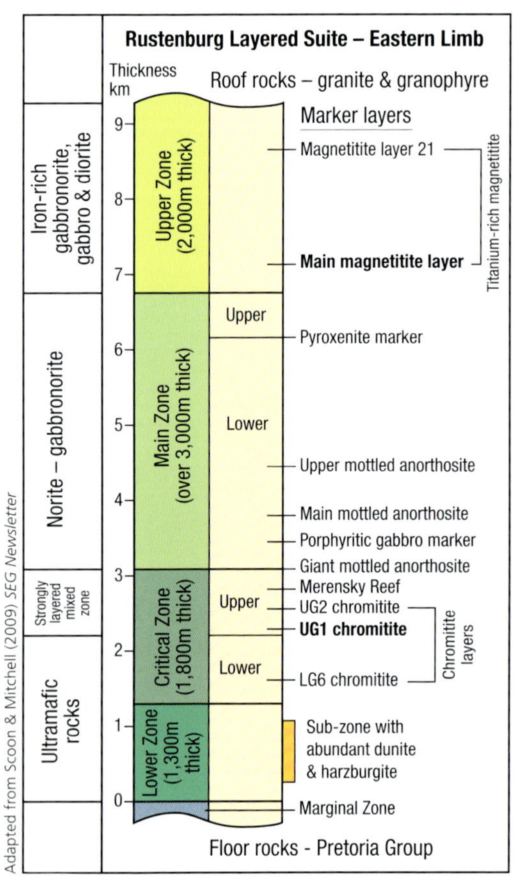

Above: A simplified cross-section of the Eastern Limb of the Bushveld Complex, showing major rock formations and the location of the Dwars River chromitite and Main magnetitite exposures.
Left: A simplified stratigraphic column through the approximately 9km-thick Eastern Limb of the Rustenburg Layered Suite of the Bushveld Complex.

Regional geology

The **Rustenburg Layered Suite** (RLS) of the **Bushveld Complex** is by far the world's largest layered mafic intrusion and contains the greatest concentration of mineral wealth on Earth, comprising platinum group metals, chromium, nickel, iron, titanium and vanadium. In the **Eastern Limb** these riches occur within an amazing 9km-thick sequence of layered igneous rocks; these are subdivided into several thick 'zones', varying from ultramafic at the base, through anorthosite, followed by norite and gabbro, which becomes iron-rich towards the top (see stratigraphic column and idealised cross-section). The magmas that formed these rocks were intruded deep within the Transvaal Basin around 2,060 million years ago.

Within the lower part of the RLS there are numerous sharply defined stratiform layers (or

Above and below: Along the minor Dwars River, the thin black UG1 chromite layers can clearly be seen on the left and are also exposed on flat surfaces below.

seams) of black, chromite-rich rock that is also called 'chromitite' and which is mined on a large scale. The chromitite occurrence found at Dwars River is one of these and is known as the 'UG1 layer' but here it is not favourable for mining. Much higher up in the RLS there is also a thick sequence of iron-rich gabbro that contains extensive stratiform layers of pure magnetite, rich in titanium and vanadium, and this is mined in places (see text box on p.213). The position of these occurrences in the RLS is shown in the diagrams opposite.

Landscape and landforms

The region is part of the strongly dissected country lying to the west of the Great Escarpment and is bisected by the northeast-trending Steelpoort Valley, a feature formed by the major Steelpoort Fault. Scenery varies from hilly to mountainous and is largely shaped by differential erosion of the gently dipping strata of the Bushveld Complex. At the site, the Dwars River – a tributary of the Steelpoort River – flows over well-exposed bedrock and through a narrow defile that is incised along a zone of jointing; here the river has cut down into the anorthosite rocks, and enhanced exposure of the chromitite seams.

On some weathered surfaces the chromite-rich layers reveal an unusual texture where they form larger clustered crystal masses. This indicates that there is a silicate mineral component that has cemented the abundant fine-grained chromite crystal clusters together.

Rock formations

- It is the striking and enigmatic relationship between thin layers of black, chromite-rich rock and thicker layers of greenish white anorthosite that resulted in this site being declared a National Monument in 1965.

- The chromitite layers consist of fine-grained chromite crystals cemented into clusters by silicate minerals, and the anorthosite consists predominantly of coarse plagioclase.
- The chromitite-anorthosite geological relationship is not easily explained as the chromite mineral grains did not crystallise in place, and scientists are still not agreed on the mechanism of formation.
- The presence of small anorthosite rock fragments within the chromitite indicate a form of horizontal injection: a chromite-rich concentration that accumulated elsewhere within the magma chamber was remobilised and intruded laterally as a 'chromite slurry' into semi-consolidated anorthosite host rocks along existing igneous layering in the anorthosite.

Above and inset: These telling photographs show small fragments of anorthosite 'floating' within the chromite layers – which means that the thin chromite layer was intruded into the earlier anorthosite layer, and is therefore younger.

This image shows that, in places, thin chromite layers are broken up within more massive anorthosite, probably explained by the latter still being in a partly plastic state after the time of chromite intrusion and solidifying.

Bushveld magnetite layers – another geological enigma

A detour off the R555 onto the road to Jane Furse (R579) will bring you to the famous Magnet Heights geological exposure near Ga-Mogashoa. It is on the east side of the road, 13km from the R579 turn-off, but regrettably there is no sign or acknowledgement of its significance. It marks another economic wonder of the Bushveld Complex: the existence of vast layers of pure magnetite rock called 'magnetitite'. Around the entire complex there are up to 25 discrete layers within the 2,000m-thick iron-rich gabbro sequence of the RLS. At Magnet Heights the approximately 2m-thick Main Magnetite layer is exposed very close to the road. The magnetitite also contains titanium (approximately 12%) and vanadium (approximately 1.6%) and it is mined on a large scale further south near Roossenekal and railed to Witbank for smelting.

The famous 2m-thick Main Magnetite layer at Magnet Heights, now named Ga-Mogashoa, close to the R579 leading to Jane Furse. Inset: Showing the very coarsely crystalline nature of this titanium- and vanadium-rich iron ore.

Things to see and do

- **Examine the Dwars River site:** A careful look at the outcrops along about a 150m stretch of the Dwars River shows the puzzling relationships described in this section. Notice the splitting of thin chromitite layers, the small anorthosite inclusions within the chromitite, the disrupted layers of chromitite in anorthosite, and the texture of the chromite rock. There is a short descriptive plaque on site.
- **See Kennedy's Vale magnetitite pipe:** This large body forms a prominent hill immediately east of the R555 and is within the Lion Ferrochrome Smelter property. It is the largest of the intrusive, carrot-shaped magnetitite pipes in the Bushveld Complex and was previously mined for its valuable vanadium content.
- **Do the Platinum Trail:** There are numerous geological sites in the Eastern Limb, which has become a major platinum and chromite mining district. To access the sites of the historic Platinum Trail you will need permission from local villagers whose land you may cross and from the local mining operators and also a geological guide book.

33 Wyllie's Poort

A PASSAGE THROUGH A ROCKY DEFILE

The only road through the Soutpansberg to the Limpopo Valley and Lowveld

The central part of the elongated Soutpansberg massif as seen from the N1 near to the Nzhele Dam turn-off, well south of Musina. In the distance you can see the start of Wyllie's Poort before it disappears into the defile.

As you near Makhado (formerly Louis Trichardt), the usually hazy blue, 180km-long mountain range of the Soutpansberg (meaning 'mountain of the salt pan') rises up impressively before you. It once proved a formidable obstacle to travel and even today the main road takes advantage of the narrow gorge, called Wyllie's Poort, that has cut down through hard quartzite.

Location and getting there

- Wyllie's Poort and its two tunnels, about 15km north of Makhado, are important landmarks on the N1. From Makhado the road climbs steeply up the southern side of the mountain before descending to the gorge of Wyllie's Poort. About 6km north of Wyllie's Poort there is a less spectacular cutting called Musekwaspoort, also traversed by the N1.
- GPS (parking at tunnels): 22°54'56"S; 29°55'41"E

Setting the scene

There are only two 'poorts' (gorges) through the east-northeast-aligned massif, one being Wyllie's Poort in the centre, well known to road travellers heading to the far north; the other is at Waterpoort in the west, which is followed by the railway line. The original and sometimes hazardous route at Wyllie's

A roadside exposure of typical hard, blocky Wyllie's Poort quartzite that dominates the pass and poort across the Soutpansberg. White lines indicate the shallow dip (about 10°) to the north of the quartzite beds (originally sandstone).

In the regional town of Makhado (formerly Louis Trichardt) and close to the N1, there now stands an impressive statue of KhosiKhulu Makhado (died 1895), who once ruled this region.

One of many north-tilted, step-like faulted blocks, seen on the mountain pass south of the gorge, dipping about 25°.

Thin bands of quartz pebble conglomerate, as seen here, are common within the Wyllie's Poort Formation quartzite and indicate that the original sandstone and conglomerate were deposited in a braided alluvial setting.

The abandoned and overgrown old road that originally went through the gorge, seen here on the left, near the entrance to the southern tunnel on the N1 to the right.

Poort was a narrow gravel road that closely followed the defile through the hard quartzite. It is named after Lieutenant CH Wyllie who surveyed and engineered the original road in 1904. Increasing traffic led to the later construction of a new road higher up, including two tunnels cut through quartzite, 457m and 274m long, opened in 1961 and known as the Hendrik Verwoerd Tunnels.

Regional geology

This unusual massif is made up of the **Soutpansberg Group**, a mixed sequence of sedimentary and volcanic formations approximately 1,900–1,700 million years old. The package is around 7,000m thick and the formations are tilted to the north at about 10–25°. Roughly east-west-trending block-faulting has resulted in a step-like repetition of the strata.

Landscape and landforms

Much of the Soutpansberg rises over 1,600m above sea level and the highest peak reaches 1,747m. This ancient range lies above the African land surface and is a dissected remnant of the surface that existed before the break-up of Gondwana. Repetition of the dipping formations has resulted in a landscape of steep scarp-slopes and more gentle dip-slopes. Cross-cutting structures have been exploited by gorges such as Wyllie's Poort and Waterpoort, the latter being eroded by the Sand River.

Rock formations

- Hard, resistant Wyllie's Poort Formation quartzite, about 1,500m thick, is the dominant geology seen along the roadside.
- Typically reddish in colour due to pervasive iron minerals, it is known geologically as a 'red bed' and is one of the oldest formations of this type in South Africa.

- It was originally deposited as sandstone on a braided alluvial flood plain, as is shown by shallow-water sedimentary structures, coarser-grained grits and pebble beds.
- But it is the presence of iron oxide as coatings around sand grains and even as thin layers along bedding planes that makes it important – the reddish quartzite represents the first iron oxide-bearing red beds in South Africa, evidence of an oxygenated atmosphere. (For an expanded explanation on this see Kloof Pass, pp.239–243.)
- Basaltic lavas that occur below the quartzite do not show well in outcrops but a thick diabase sill is exposed near the southern tunnel.

Another rugged, block-faulted hill of north-dipping, strongly jointed Wyllie's Poort quartzite near the southern tunnel entrance.

At the Hendrik Verwoerd Tunnels parking area there are fine exposures of fresh Wyllie's Poort red bed quartzite. The rock is pink to maroon in colour thanks to its high iron content. The dark grey mineral is hematite that has recrystallised into thin layers and veinlets.

An exposure of the same pebbly, hematite-rich Wyllie's Poort red bed quartzite seen at Masekwaspoort, about 6km from the northern end of the gorge.

The Soutpansberg graben

The Soutpansberg massif was formed in a large east-northeast-trending graben structure (a dropped-down block of land bordered by parallel normal faults) that became filled mostly by basaltic volcanics and sandstone, the latter becoming hard Wyllie's Poort quartzite. The origin of the unique graben-like geological setting is still being debated. On both its southern and northern sides the Soutpansberg range is bounded by major fault structures against older granitic gneisses of the Limpopo Complex. Many strike-faults cut across the range in much the same direction and have resulted in the step-like block-faulting. Differential erosion between more easily weathered granitic gneisses and harder quartzite resulted in the high-standing residual massif, but in the north the contact is now covered by younger Karoo formations.

There are two possible explanations for this ancient graben: 1) the rocks of the Soutpansberg were deposited within a fault-bounded, 50km-wide rift valley, akin to the East African Rift Valley we see today; or 2) the rocks were deposited in a much broader basin that covered a larger part of the Kaapvaal Craton and were later preserved only in a wide graben-like structure. Perhaps the answer lies in a combination of these ideas, the lower formations being confined within the early graben structure and the upper formations within a broader basin, since lost by erosion. In a geological paradox, it now forms a remarkable mountain range.

Things to see and do

- **Notice the faulting:** The step-like topography seen on the mountain road reflects discrete, north-dipping fault-blocks. Stop before the entrance to the southern tunnel and notice the highly jointed rock face above the tunnel portal, manifestation of a large fractured fault-plane, and one of many such faults that exist in the area.
- **Stop between the tunnels:** From the parking area you can see the narrow poort and the old road (now closed to traffic) that was the main artery for over 50 years. Have a look at the hard, iron-rich red bed geology in Wyllie's Poort quartzite in the road cutting nearby. Similar rocks can be seen at Masekwaspoort further north.

Two tunnels have been blasted through the solid Wyllie's Poort quartzite. On the left is the southern tunnel where an almost vertical fault-scarp is exposed, revealing highly jointed quartzite. The northern tunnel is on the right and Wyllie's Poort gorge lies to the right of the tunnel.

34 Sand River Gneiss

WORLD-RENOWNED GEOLOGICAL SITE

Rocks that reveal the titanic crustal clash between the Zimbabwe and Kaapvaal Cratons

Part of the spectacular gneiss and migmatite pavement that is exposed in the rocky, largely unweathered and usually dry bed of the Sand River.

The road to Musina crosses an unspectacular plain before dropping down into the dissected terrain of the Limpopo River Valley. The usually dry sandy landscape of scrubby mopane trees and baobabs does no justice to the primeval Earth-making drama that once took place here, in what is, geologically speaking, the Limpopo Belt.

Location and getting there

- Musina (formerly Messina) is at the 'top' end of the N1 and is South Africa's most northerly town. The world-renowned outcrops of Sand River gneiss lie about 9km to the southeast, close to the Tshipise road (R508) at the old concrete Causeway that crosses the Sand River. This is within the Musina Nature Reserve and access can be gained only through the main entrance on the N1 about 4km south of Musina, where you will get directions to the usually dry river bed.

- GPS (Causeway area, off R508): 22°24′12″S; 30°05′57″E

The Sand River, seen from the R508 road bridge, viewed looking west. This site can be accessed only via the Musina Nature Reserve off the N1. The old Causeway, which crosses some of the most spectacular geology, is indicated by the dashed white line. This is often called the 'Causeway locality'.

A closer look at the incredible plastic deformation of light-coloured Leucocratic gneiss and darker Grey gneiss to form an extremely complex mixed rock called 'migmatite' – a rock consisting of an intimate mixture of at least two other rocks.

Setting the scene

Early on in Earth's history, during the Archaean, two small continental nuclei started drifting towards each other, part of the movement of tectonic plates. These were what we now call the Zimbabwe and Kaapvaal Cratons, and their collision may have created the world's first continent. This is viewed by some geoscientists as the world's oldest continent-making event, and it had far-reaching consequences for the geological development of southern Africa. The highly contorted and strongly metamorphosed formations are the roots of one of Earth's earliest mountain ranges. The most spectacular outcrops occur along the bed of the Sand River where the exceptional geology is wonderfully exposed on the scoured rock pavement.

Regional geology

The rock formation is known as the **Sand River Gneiss**, a distinct geological unit found within the extremely complex **Central Zone** of the **Limpopo Belt**. Rock-dating has given ages of around 3,400–3,200 million years, revealing that these gneisses are the oldest rocks of the belt. Current geological thinking is that they represent the ancient basement onto which the younger supracrustal rocks of the **Beit Bridge Complex** were deposited, but not all geologists agree with this theory.

Landscape and landforms

The Sand River is a major tributary to the Limpopo River, the confluence lying about 10km downstream from the old Causeway, and the area is part of the now dissected post-African land surface. Locally there are impressive examples of exfoliating granite-gneiss domes. But most revealing is how, over millennia, the strongly abrasive action of the Sand River has scoured the gneisses to produce fresh rock surfaces ideal for detailed geological study.

Rock formations

- These gneisses are called 'migmatites' (consisting of an intimate mixture of at least two other rocks) and show extreme metamorphism and plastic deformation. They are of magmatic origin (orthogneiss), formed from some earlier granitic rocks.
- They are strongly banded, folded in a complex way and show amazing stretching, interlayering and contortion.

● These rocks have been the subject of considerable geological research for many years and several episodes of strong deformation have been revealed.

● Also present are felsic dykes that represent younger granitic intrusions, and strongly deformed amphibolites that were orignally mafic dykes.

● Much younger dolerite dykes cut through the gneisses without deformation, indicating their much later intrusion.

The dark rocks shown here are thought to represent a much earlier mafic dyke that intruded the gneiss and which has subsequently been intensely deformed.

An elliptical and steeply plunging sheath fold within Leucocratic gneiss and darker Grey gneiss, the result of complete detachment of an elongated cylindrical structure during intense deformation.

Scattered patches of lime-green epidote and pink potassic feldspar reflect hydrothermal alteration that took place long after the intense folding.

What is the Limpopo Belt?

This is a broad zone of gneisses and metasedimentary rocks characterised by intense structural deformation, strong metamorphism and melting of the original rocks during several major tectonic episodes 3,300–2,000 million years ago. The Central Zone, as represented at the Sand River, is the most intensely metamorphosed and structurally deformed part. It is up to 160km wide and consists of a wide variety of gneissic rock types that show no uniform structural pattern. While the full story of this very complex geological terrain is by no means entirely understood, we know that, broadly speaking, the following took place:

● As the Zimbabwe and Kaapvaal Cratons moved closer together, the thin crust between them became compressed, strongly folded and thickened to a depth of more than 30km, forming a range of Himalayan proportions.

● Deep within the crust these rocks were intensely metamorphosed and partially melted to form a range of granitic magmas that intruded the crust, and eventually became the Sand River gneisses.

● The process of mountain building formed an elongated 'Limpopo' mountain range, and also welded the cratons together to make what was probably Earth's first continental mass.

● Over geological time the mountains were completely eroded so that today nothing is left but their twisted and confusing roots. However, the full geological story of this structurally complex terrain has not yet been entirely unravelled.

Things to see and do

Such is the geological importance of the old Causeway locality that at one time it was mooted to become a 'National Monument'. This did not happen but you will be able to see a wide variety of gneisses and intrusive rocks, of which the major rock types are listed below and some of which are also illustrated.

- **Leucocratic gneiss and Grey gneiss:** These two orthogneiss units are complexly interlayered and tightly in-folded. They underlie most of the area to the south of the old Causeway and were originally of granodiorite and quartz diorite composition.
- **Dark amphibolite-rich gneiss:** Only seen north of the old Causeway and forms a distinct unit known as the Beit Bridge Complex, representing a paragneiss complex of meta-sedimentary rocks.
- **Felsic dykes:** Not easily recognised within the Leucocratic or Grey gneiss; these dykes are irregular, light-coloured intrusions that are in-folded into the gneiss.
- **Mafic dykes now metamorphosed to amphibolite:** Highly irregular, dark grey, complexly in-folded and dismembered, they include the Stockford dykes about 1.5km upstream.
- **Pegmatite:** Irregular, whitish, coarse-grained intrusions, usually vein-like and consisting mainly of feldspar; clearly younger than some of the gneisses.
- **Dolerite dykes:** Dark brown to black, fine-grained, vertical and undeformed, they impressively cross-cut all the other rocks.

A large boulder of migmatite showing highly complex folding of different types of gneiss.

An example of 'boudinage' where, during intense deformation, the more competent (resistant to deformation) light-coloured gneiss has been stretched within more plastic Grey gneiss, resulting in 'pinch and swell' structures.

A relatively young, narrow and undeformed dolerite dyke that cuts cleanly through the migmatitic gneisses and clearly postdates their formation.

35

Mapungubwe

WORLD HERITAGE SITE

South Africa's most celebrated archaeological site

Courtesy of University of Pretoria (UP)

The small mesa known as Mapungubwe Hill, seen from its western side after summer rain. Once a royal citadel, the wedge-shaped hill is about 320m long and stands up to 40m above the sandy valley. With sheer sandstone cliffs all around, the only access now is via a wooden stairway on the northern side.

Mapungubwe, once the capital city of a sophisticated African Iron Age kingdom and centre of a thriving farming, hunting and trading society, became the focus of major archaeological research that started in the 1930s and continues to this day. Starting around AD 900, prehistoric societies prospered in the Limpopo River Valley for nearly 400 years, and had trading links with the East. Around AD 1300 the city-state went into rapid decline and was abandoned, possibly for climatic reasons, and the inhabitants dispersed.

Location and getting there

- The World Heritage Site is located in the far north of Limpopo province, near the confluence of the Limpopo and Shashe Rivers where Botswana, South Africa and Zimbabwe meet, and within what is now Mapungubwe National Park. Get there either via Musina and then drive west along the R572; or via Alldays on the R521, then head northeast on the R572.
- GPS (SANParks entrance gate): 22°14′37″S; 29°24′03″E

Setting the scene

On 8 April 1933 *The Illustrated London News* announced: '… a remarkable discovery in the Transvaal: a grave of unknown origin, containing much gold work, found on the summit of a natural rock stronghold in a wild region'. For countless generations local people had held Mapungubwe Hill in considerable awe and reverence, and whatever secrets it held remained undisturbed. But there were rumours of riches and on 31 December 1932 a local teacher, Ernst van Graan, together with his son and friends, reached the prehistoric burial site on top of Mapungubwe Hill. They were astonished to find clay pottery, iron tools, copper wire, glass beads, gold beads and plate, and human skeletal remains. The University of Pretoria was informed, then the government became involved, followed by intensive and ongoing meticulous archaeological exploration and research. It was declared a World Heritage Site in July 2003 and is the core of the Greater Mapungubwe Transfrontier Conservation Park.

Courtesy of University of Pretoria (UP)

An early archaeological study by the University of Pretoria at the K8 excavation site on Mapungubwe's Southern Terrace. This is now the site of a very effective below-ground display.

Roger de la Harpe / Africa Imagery

An undated aerial view of Mapungubwe Hill, seen as one faces westwards. The Southern Terrace is on the far side of the hill and shows the K8 excavation shelter. Archaeological excavations on the top of the hill are clearly visible.

Remnants and collapsed boulders of Tshipise (Clarens) sandstone overlying greyish Red Rocks (Elliot) siltstone can be seen on the way to Mapungubwe Hill.

Regional geology

The area forms part of the outlying Tuli Basin, a narrow east-west sedimentary basin of Karoo age that comprises a restricted sequence of the **Karoo Supergroup**. In Mapungubwe National Park only the **Clarens Formation** is exposed, consisting of upper **Tshipise Member** sandstone overlying **Red Rocks Member** siltstone. In the Tuli Basin the Tshipise sandstone is the equivalent of the well-known Clarens Formation, while the underlying Red Rocks siltstone probably represents the top of the **Elliot Formation**.

Landscape and landforms

Erosion of the sandstone has formed a landscape of numerous small, steep-sided mesas and isolated buttes separated by wide sandy valleys, eroding towards the Limpopo River, and creating a minor sandstone escarpment at the edge of the plateau. Some of the mesas are protected by a thin layer of lava. Contrasting rates of erosion between the sandstone and siltstone under semi-

Large-scale aeolian cross-bedding (the planes shown by the white lines) reveals the windblown origins of this Tshipise (Clarens) sandstone (a fossil dune), which overlies pale greyish Red Rocks siltstone (below yellow line). 'Elephant-skin' cracking on the surface is a weathering effect.

South Africa's first gold workers

Mapungubwe is renowned for its numerous 13th century gold-plated artefacts, including the iconic Gold Rhino. These were found alongside human remains at the hilltop burial site, and are thought to have had strong ceremonial and religious value. The goldsmiths were skilled, able to produce thinly beaten gold plate, and making thousands of gold beads and minute gold nails. These are the earliest known gold artefacts found in southern Africa, and indicate that gold production was well established. Certainly, the source of the gold was not from around Mapungubwe, but easily worked alluvial gold was plentiful further north in today's Zimbabwe, where over 4,000 pre-colonial gold mines have been discovered.

The iconic Gold Rhino that was excavated at Mapungubwe Hill. Thinly beaten gold plate was tacked onto a wooden base, using small gold nails. Interestingly, it does not have the second, smaller horn typical of the African rhino.

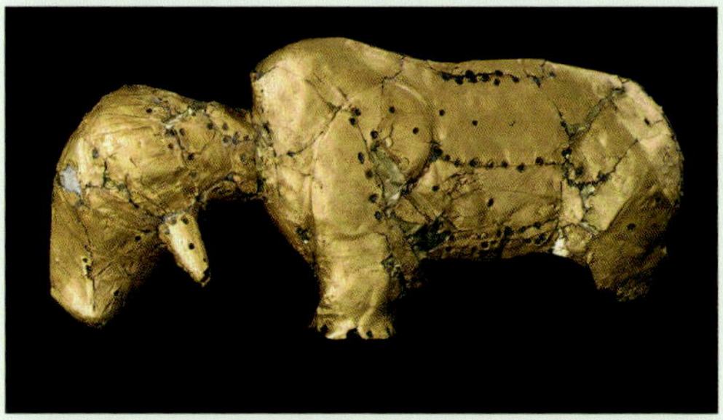

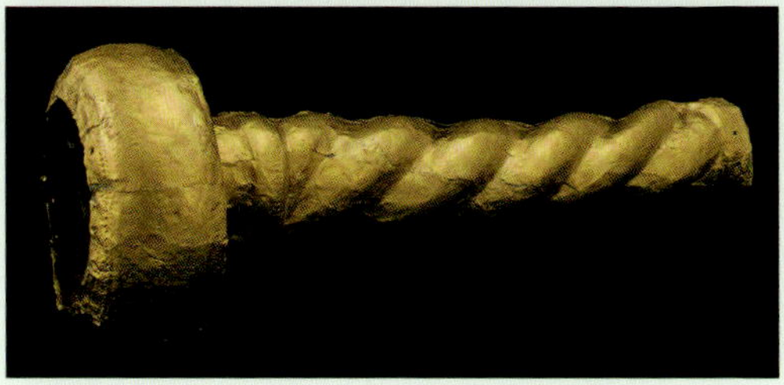

Numerous other priceless gold artefacts were excavated with the human remains at Mapungubwe Hill, including bovine and feline figurines, a sceptre and a gold bowl.

All images courtesy of University of Pretoria

arid conditions have resulted in spectacular landforms, assisted by the pronounced pattern of northwest-trending jointing. This has resulted in a 'badlands' landscape, enhanced by the sandstone weathering to attractive golden-brown colours. Driving through the sandy valleys you get to see how mesas and buttes, and sandstone caves and overhangs are formed by denudation of the old land surface.

Rock formations

- Tshipise sandstone: This spectacular, fine-grained, whitish sandstone is up to several tens of metres thick and is of windblown (aeolian) origin, revealing the desert conditions that existed here around 190 million years ago. This was the final phase of Karoo sedimentation.

Things to see and do

- **Mapungubwe Hill heritage tour:** To get there you need to do the official guided tour that takes you to the top of the iconic sandstone citadel through a cleft in the rocks. At the base of the hill on the Southern Terrace is the below-ground K8 archaeological exhibit that graphically reveals evidence of past habitation in the area.
- **Drive the park:** A good gravel road goes all the way around the eastern section of the Park and provides exceptional views of the rivers, geology and landforms:
    - See how Tshipise sandstone has weathered and eroded, and look out for the aeolian cross-bedding and peculiar 'polygonal' weathering structures.
    - View the contact zone between upper sandstone and lower siltstone, a disconformity surface that represents an old land surface and a geological time gap (of unknown length) before the desert sand was deposited.
    - From the Pinnacle viewsite see the clusters of small sandstone pillars that are interpreted as fossilised termite nests. These completely lithified structures, showing internal tubes and passages, represent about 190-million-year-old termitaries that existed in moist interdune locations. These are the best-known examples of fossilised termitaries in South Africa, and perhaps the oldest ones in the world.
- **Visit the Interpretation Centre:** This architecturally impressive, prize-winning construction is located near the main entrance. Many archaeological treasures have been returned to their natural home here and are on display. If you cannot get to this site, there's also an astonishing display of artefacts at the Mapungubwe Collection in the Old Arts Building at the University of Pretoria – another must-do visit.

Looking up the wooden steps that lead through a cleft in the massive Tshipise sandstone – the only way to the top of Mapungubwe Hill.

Mapungubwe's new, award-winning Interpretation Centre, made of local stone, houses a fine collection of treasures.

● Red Rocks siltstone: Below the sandstone is pale greenish-grey to reddish siltstone reflecting silt and mud deposited on a semi-arid river flood plain. The siltstone immediately below the sharp contact with the sandstone was once an exposed ancient land surface, before desert conditions took over.

● Letaba basalt: Karoo sedimentation was followed across much of southern Africa by Karoo continental volcanism (see Amphitheatre, pp.128–133 and Sani Pass, pp.123–127). At Mapungubwe all that is left of this are now isolated remnants of the dark, fine-grained lava that caps some of the hilltops.

● Dolerite dykes: Several coarsely porphyritic dolerite dykes, part of a major dyke swarm, were intruded later. In places, these dykes are reminiscent of man-made walls.

*The very sharp contact between massive Tshipise (Clarens) sandstone (**A**) and underlying Red Rocks (Elliot) siltstone (**B**), which formed an ancient weathered land surface before being covered by desert sand around 190 million years ago (see p.147 for an artist's impression).*

The unusual porphyritic and coarse-grained texture of a dolerite dyke that cuts through sandstone near the Interpretation Centre. This is one of a swarm of east-west-trending Karoo (Letaba) dolerite dykes in the region.

Emese Bordy

A cluster of rock-hard fossil termite nests (around 190 million years old) seen on the hillside south of the Pinnacle viewsite near the confluence of the Limpopo and Shashe Rivers. They are embedded in Tshipise (Clarens) sandstone, contain complex networks of sandstone tubes and cavities, and can exceed 2m in height.

36 Eersteling Gold Mine

PROVINCIAL HERITAGE SITE

Where South Africa's illustrious gold-mining industry first started in 1871

Museum Africa, City of Joburg

By an unknown artist, this watercolour painting depicts the Eersteling Gold Mine around 1880 (viewed from the south); the chimney and boiler-house are clearly evident.

The historic 16m-high, stone-built chimney and the remains of a boiler-house built around 1872, using local stone. In the distance, to the southwest, are the Strydpoort Mountains which are topped by Black Reef quartzite overlying formations of the Wolkberg Group at the base of the Transvaal Basin.

After spending several years travelling around southern Africa, German explorer and amateur geologist Karl Mauch predicted a rich golden future for the land, and his enthusiastic reports attracted many fortune hunters to the old Transvaal Republic and beyond. Among them was Edward Button, an energetic prospector from Natal who would go on to make mining history.

Location and getting there

- The Eersteling goldfield lies about 25km northeast of Mokopane (formerly Potgietersrus). To get there, take the old road to Polokwane (formerly Pietersburg) and turn off at the sign 'Eersteling Monument' about 18km from Mokopane. This gravel road takes you to the old stone chimney about 5.5km from the R101.
- GPS (chimney parking area): 24°07'32"S; 29°16'13"E

Setting the scene

In 1871 Edward Button, an experienced gold prospector, found himself in John Goodwin's trading store at the settlement of Marabastad near present-day Polokwane. There he met Jacobus F du Preez, who showed him samples of gold-bearing quartz that came from his farm Eersteling, and who invited Button to prospect on the farm. Button soon found rich alluvial gold coming from a deeply weathered reef. Button and Goodwin then bought the farm for £50 and named the rich gold strike the Natalia Reef. Alluvial gold was recovered by sluicing, but soon gold was being won by crushing weathered quartz reef in primitive see-saw fashion under a large round boulder, devised by William Pigg, one of Button's mining partners. The site was declared a National Monument in 1938 and is now a Provincial Heritage Site.

Port Elizabeth Public Library

Artist Thomas Baines visited Eersteling in late 1871, recording this impression of men crushing gold ore using a see-saw on a large round boulder, then panning the resulting rock powder to recover the gold – a method devised by William Pigg, one of Button's partners.

Regional geology

The goldfield lies at the southwestern end of the **Pietersburg Greenstone Belt** in metamorphosed volcanic rocks known as the **Pietersburg Group**. This is a narrow belt of northeast-trending Archaean greenstones surrounded by intrusive **Basement** gneisses and granitic rocks, probably less than 3,000 million years old, but similar to formations found in the **Barberton Mountain Land**.

Landscape and landforms

East-west-trending low hills are formed by harder basalt, while valleys are underlain by softer mafic schist and granite. To the south and west, the Strydpoort Mountains, made of hard Wolkberg sedimentary formations and topped by resistant Black Reef quartzite, form a major escarpment that is a remnant of an older pre-African erosion land surface. The highest peak, at 1,953m, is fittingly called Button's Kop.

Rock formations

- The dominant rocks at the old mine are mafic volcanics (basalt) of the Eersteling Formation, now metamorphosed to a dark, fine-grained amphibolite and strongly sheared in places.
- A careful look at some of the basalt outcrops shows evidence for pillow lavas, telling that the molten lava erupted and solidified under the sea as part of Earth's early crust.

Gold ore

As is typical of greenstone belts, the ore occurs mostly in quartz-carbonate-sulphide veins, also called 'reefs' that formed when hot, mineral-carrying solutions penetrated into cracks caused by shearing. Close to the surface, the ore would have been both weathered and enriched in gold, making mining and gold recovery easier, and much of it would have found its way into the streams as alluvial gold. There are several reef zones, which have been given names such as the Doreen, Pienaar's and Girlie reefs.

Thought to be the original steam boiler that was imported from the United Kingdom, this mining heritage object now lies abandoned near the old chimney site.

The wall of the chimney bearing an assortment of engraved names. On the right is 'W. Pigg 1872'; William Pigg, who arrived that year, was one of Edward Button's mining partners.

This is the original boulder-crushing site on Natalia Spruit, as seen in the Thomas Baines painting on p.231. The bowl-shaped site is on a flat ledge of basalt.

South Africa's first gold mine

In September 1871 Edward Button appeared before the Volksraad in Pretoria to announce the discovery of a profitable goldfield. In December that year the Marabastad goldfield was proclaimed, the first in South Africa, and prospectors flooded in. Greatly enthused by their find, Button and Goodwin set off for London in 1872 where they raised £50,000 and formed the Transvaal Gold Mining Company Limited, South Africa's first gold-mining company. At first the ore was crushed under a rocking granite boulder; but soon a 12-stamp rock crusher, steam engine and boiler were purchased, shipped to South Africa and brought to Marabastad by ox-wagon – at great expense. The crusher came into operation only in 1874 and early production results are unknown; but in January 1875 a newspaper reported that 11.665kg had been recovered in the latter part of 1874. The mine operated until the start of the 1st Anglo-Boer War (1880–1881) and resumed in 1885, when prospecting in the general area also increased. From 1906–1937 the Eersteling goldfield is reported to have produced 439kg of reef and alluvial gold, and over the years numerous small workers continued to mine in the area. However, the richer goldfields at Pilgrim's Rest, Barberton and the Witwatersrand soon eclipsed this small mine.

The steel headgear of the Franka Shaft, part of the modern-day Eersteling Gold Mine, which started mining again in 1987 but is now closed down.

Things to see and do

- **See the old chimney and boiler-house:** The 16m-high chimney is built of dressed local stone of various types. Look out for historic names engraved in the rock wall – including that of 'W. Pigg 1872', who ran the operation in Button's absence.
- **Look for the original gold crushing site** in nearby Natalia Spruit and see the basalt host rocks.

Unweathered, dark massive basalt, metamorphosed to fine-grained amphibolite, exposed in Natalia Spruit. In places these rocks are also strongly sheared.

37 Makapans Valley

PART OF A WORLD HERITAGE SITE

This valley has one of the world's greatest fossil records of human evolution

Looking northeast up the Makapans Valley from the Cave of Hearths. The resistant, cliff-forming formation in the distance is of westward-dipping Black Reef quartzite; this underlies the thick Malmani dolomite sequence in the valley, in which the many caves have formed.

The many fossil-rich dolomitic caves in Makapans Valley have a history that goes back to hominid species around 3 million years ago, and they provide a continuous record of human habitation from the Early Stone Age to recent times. They are linked to the Cradle of Humankind World Heritage Site and are of international scientific importance.

Location and getting there

- Mokopane (formerly Potgietersrus) is the starting point for a visit to Makapans Valley. Take the old main road (R101) leading to Polokwane (formerly Pietersburg); about 13km from the town centre turn right onto a signposted gravel road and drive a further 7km to the old Limeworks Cave, where you will meet a local guide (by prior arrangement).
- GPS (gate into Limeworks Cave): 24°08′22″S; 29°11′23″E

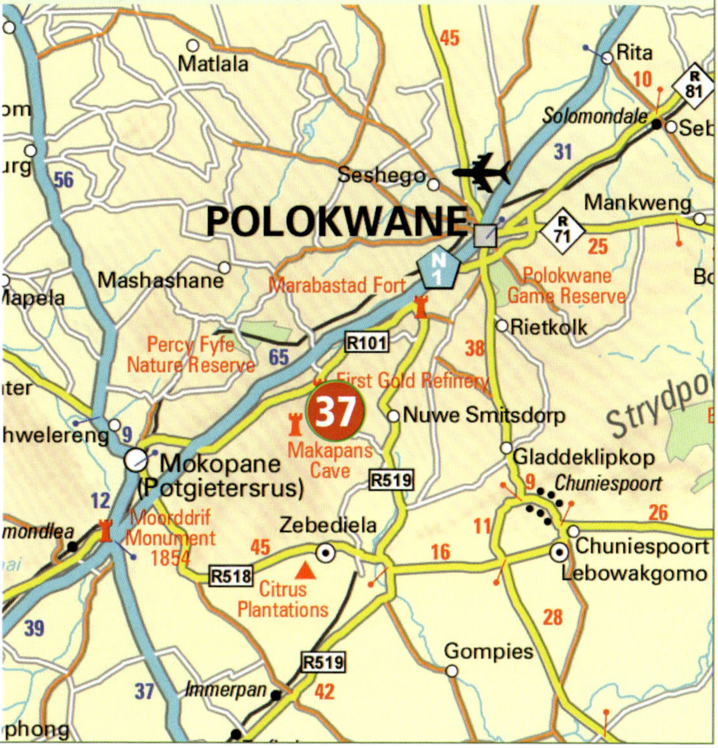

Setting the scene

Makapans Valley is the site of numerous caves of scientific and historic interest and an outlying part of the Cradle of Humankind World Heritage Site. Most caves are not accessible to the public.

- Limeworks Cave is a treasure-house of fossils. Lime-workers, digging out pure cave limestone to burn for quicklime, removed masses of fossil-bearing cave breccia. In 1925 a local teacher, Wilfred Eitzman, became aware of the fossil legacy and reported this to Professor Raymond Dart, thus securing the debris for later study.
- Further up the valley is the Cave of Hearths, famous for evidence it presents of hominids and Stone Age man. It has been the focus of academic research for many years.
- Elevated above this is the original Makapans Cave, now known as Historic Cave or the Cave of Gwasha. In 1854 a large number of Kekana

Limestone miners at Limeworks Cave removed large amounts of secondary cave breccia in order to access the limestone (travertine) cave formations. Later collected by researchers, the debris was stacked in rows in the 'cave breccia yard'. Almost every piece of rock contains at least one fossil.

This is a typical example of grey bedded Malmani dolomite exposed outside Limeworks Cave. The more prominent, lighter-coloured bands are bedding-parallel chert layers.

An outcrop of characteristic 'elephant-skin' weathering of massive Malmani dolomite at Limeworks Cave.

Ndebele under Chief Mokopane took refuge here to escape reprisal after being associated with a Voortrekker massacre. They were besieged by a large Boer commando and after 30 days many had succumbed to thirst and starvation, and many were killed. It was listed as a National Monument in 1936.

Regional geology

The valley is carved into the **Malmani Subgroup**, a dolomite sequence that is more than 1,000m thick, and is part of the **Transvaal Supergroup**. Located along the northern rim of the Transvaal Basin, these formations are faulted and dip mainly to the southwest. The mountains to the northeast are capped by resistant **Black Reef Formation** quartzite.

Landscape and landforms

The surrounding high ground is a remnant of the land surface that predated the African cycle of erosion, but it is now generally dissected and lowered. The caves formed very much later, probably during a time of wetter climate in the Tertiary. Their formation resulted from a chemical weathering process called 'karsting', a common development in limestone and dolomite (see Cango Caves, p.84).

The entrance to Limeworks Cave: an official guide will take you through the large cave system, and there are excellent explanatory signboards outside.

Rock formations

- Throughout the valley the local rock is dark grey, fine-grained Malmani dolomite, a rock made largely of calcium-magnesium carbonate. The original sediment was

When environmental conditions are right, cave formations continue to grow, even today. These young formations on an inside wall of Limeworks Cave could be the start of another drip-stone formation.

A close-up view of unidentified fossil bone breccia from the 'cave breccia yard'.

deposited in the shallow Transvaal Sea around 2,600–2,500 million years ago.

- Deposition took place during a geologically very quiet period when the dominant life form was blue-green algae (cyanobacteria), which formed vast marine mat-like algal layers and small domal structures. Today we see these as fine algal laminations and wavy stromatolite beds.
- The algae extracted carbon dioxide from sea water by means of photosynthesis, releasing oxygen into the sea water and chemically precipitating fine calcium carbonate.
- Only many millions of years later were the travertine cave formations created during wetter climatic times in the Tertiary.

Fossil and archaeological heritage

In 1925, having discovered abundant fossil bones in dumps of waste rock discarded by the lime-workers, local teacher Wilfred Eitzman sent fossil-bearing breccia to Professor Raymond Dart at the University of the Witwatersrand, who recognised the importance of the fossil bones. This focused attention on the area and many well-known scientists arrived to excavate the caves. In 1947 more systematic work was carried out and hominid bones were discovered, and explorations were extended to a number of neighbouring caves.

Cave breccia at Limeworks Cave was carefully hand sorted to recover the enormous number of fossil bones that had accumulated over millions of years. During Pleistocene times Makapans Valley was home to a great diversity of animal life, including extinct buffalo, pigs, horses, monkeys, antelope and carnivores. In the Cave of Hearths a number of 3-million-year-old hominid (apeman) specimens were also found. There is excellent evidence of a variety of Stone Age tools here, as well as evidence of more recent Iron Age occupation.

The enigmatic 'Makapansgat pebble'

Now in the Natural History Museum, London, this reddish jasper pebble, weighing 260g, was found in 1926 by Wilfred Eitzman in the Cave of Hearths during excavation of hominid fossils. It has on it natural wear patterns crudely resembling human faces. The pebble does not originate in the immediate area, and it's thought that an apeman (hominid) species could have recognised it as showing symbolic faces and brought it back 'home' in an act of aesthetic 'human' thinking. It could be the oldest art object in the world!

Things to see and do

- **Visit the Arend Dieperink Museum:** Located in Mokopane at the historic Klipskool – the old stone-built school, now converted into a museum – there is an exhibit, 'Makapansgat: Caves through the Ages', featuring the history of the nearby valley, which makes it a good place to start before going to see the caves.
- **Go to Limeworks Cave:** This is the most accessible of the caves, where thousands of fossils from a wide range of species have been identified and catalogued, including several apeman (hominid) specimens. Discarded fossil-bearing cave breccia was brought to the attention of Professor Raymond Dart in 1925.
- **See the Cave of Hearths:** Further up the valley, this cave was discovered in 1937 and named after sites of apparent ancient fires – an idea that is now discredited. But there is evidence of a continuous sequence of stone artefacts (tools) from this cave, from Early Stone Age, through Iron Age, to the arrival of the Voortrekkers.
- **Visit Historic Cave:** This vast cavern is actually the collapsed roof above and adjacent to the underlying Cave of Hearths. It was in this cave that Chief Mokopane's people took refuge in 1854 and you can still see the stone-wall fortifications. Information on the sites is well presented outside all the caves.

In weathered dolomite, this is a remnant of finely banded cave formation, also called 'travertine', which was not extracted by the lime-workers.

The famous Cave of Hearths, once thought to be the site of ancient fireplaces, and where considerable archaeological work has been carried out. Here a painted time line on the cave wall indicates age levels of archaeologically excavated sediment that had accumulated over thousands of millennia. Stone pillars have been built to support the overhanging Malmani dolomite.

Above the Cave of Hearths is Historic Cave, the large, younger cavern where Chief Mokopane and his people took refuge from the avenging Boers in 1854.

38 Waterberg's Kloof Pass

WATERBERG BIOSPHERE RESERVE

A thick sequence of red bed sandstones that reveal an important geological and environmental marker in the evolution of the Earth

Near Mamatlakala village, at the eastern start to Kloof Pass, which continues up the deep, verdant valley to the Waterberg Plateau, seen in the far distance. The development of the kloof has been assisted by a linear east-west-trending geological structure.

Road access to the high tableland of the Waterberg is limited to five natural passes. One on the eastern side is Kloof Pass where a new road was built in the 1980s, following the route of a minor track, and so opening up the internal geological structure of this part of the Waterberg.

Location and getting there

- Kloof Pass lies some 50km west of Mokopane (formerly Potgietersrus) on the steep southern side of a narrow valley (or kloof) incised by the Mmetlane Stream, a tributary of the Mogalakwena River (formerly called the Sterk River). The pass is on the R518, the tarred road between Marken on the Waterberg Plateau and Mokopane, and is about 5km long. The village of Mamatlakala is at its eastern end.
- GPS (bridge on pass): 23°58′33″S; 28°32′08″E

Looking eastwards from near the top of Kloof Pass towards the Mogalakwena River, with the plains of the Bushveld in the far distance. All the hills are of Waterberg sandstone, which is well exposed along the roadside.

Setting the scene

Completion of the new road up Kloof Pass in 1988 was a major feat of modern road-building. The numerous freshly exposed rock faces provide a magnificent view of primary depositional sedimentary structures and secondary diagenetic changes within the sandstones. Diagenesis is the process of chemical and physical change in deposited sediment during its conversion to rock. Their greyish brown to maroon mottled coloration is due to abundant iron oxides, in places further oxidised to rusty pigmentation. These are what geologists call red beds and it is here that we see early evidence of the presence of free oxygen in the atmosphere (see text box on p.243).

Regional geology

Most of the **Waterberg Group** occurs in the fault-bounded Main Basin where it now exists as an extensive topographic feature known as the Waterberg or Palala Plateau. The group comprises seven formations of horizontal and trough cross-bedded sandstones and conglomerates that are predominantly of braided river origin, and lesser formations that were deposited under both lake and desert conditions. Sediment was eroded from the ancient Limpopo highlands in the northeast and was deposited on a flood plain around 1,900–1,800 million years ago. The **Mogalakwena Formation**, less than 1,500m thick, is well displayed on the pass, but underlying this is the start of the older **Makgabeng Formation**.

Landscape and landforms

The landscape is part of the prominent Waterberg Mountain Escarpment, which is eroded along the southern and eastern margin of the Waterberg Plateau, and is locally known as the Sandriviersberge and the Hanglipberge. To the south of Kloof Pass the magnificent, 1,760m-high Hanglip buttress overlooking the Entabeni Conservancy is one of the Waterberg's iconic scenes. The Kloof Pass Valley has cut back into the plateau, assisted by some undefined structure – a possible fault – that shows on the geological

Mamatlakala Mountain (1,279m) is an isolated mesa at the eastern end of the pass that has impressive sandstone cliffs and overlooks the Mogalakwena River. The horizontal sedimentary layering is conspicuous.

Waterberg sandstone showing the conspicuous, maroon-coloured concentrations of fine hematite within paler pinkish pebbly sandstone. The hematite-rich sandstone occurs irregularly, mainly along bedding surfaces.

Part of a cross-bedded fluvial channel structure in Waterberg sandstone, an indicator of the shallow-water origins of this geological grouping.

This exposure shows typical shallow-water, channel-like cross-bedding and scattered quartz pebbles, indicating deposition on a braided, alluvial plain.

The tenacious roots of the rock fig (Ficus abutilifolia) *have established a good hold on exposed Waterberg sandstone along the roadside.*

In places rock falls from the sidewalls are not uncommon, especially after heavy rain, and this sometimes results in the pass being closed.

map as a northwest-trending lineament. On the northern side of the kloof the sandstone reaches an elevation of 1,463m, while to the southeast the landscape is dominated by an isolated sandstone mesa called Mamatlakala, which shows flat sedimentary bedding.

Rock formations

- Superb examples of horizontally bedded sandstones – exposed by blasting during the construction of Kloof Pass – reveal that the Mogalakwena Formation consists entirely of maroon to grey-brown, coarse- to medium-grained, quartz-rich sandstone and occasional thin conglomerate layers and scattered pebbles.
- These rocks were deposited on extensive flood plains crossed by multiple braided river channels that rapidly transported an abundant supply of sandy sediment from highlands located to the northeast to a shallow sea in the southwest.
- There was no vegetation on land at that time although there is probable evidence elsewhere that considerable microbial life existed on land.
- These rocks contain abundant iron oxide in the form of hematite that occurs as coatings around sand grains or as thin layers along bedding planes or cross-beds, and this feature is an important geological marker in Earth's evolution.

The world's earliest red beds

Red beds are sedimentary rocks, usually sandstones, that are reddish in colour due to the presence of iron oxides that typically coat the sand grains. Those seen in Kloof Pass are some of the oldest red beds on Earth, and it is the presence of dark iron oxide that makes these rocks important. Before 2,000 million years ago the atmosphere contained no free oxygen but, by at least 1,900 million years ago, oxygen (derived by photosynthesis from blue-green algae) was gradually becoming available in the air. On the Earth's increasingly weathered surface oxygen began reacting with reduced (oxygen-poor) iron-rich minerals such as magnetite and pyrite, converting them to rusty iron-rich hydroxides such as limonite and iron-rich clays, which were deposited within the sand, also staining and cementing it. During burial and the process of rock formation, the mobile iron hydroxides were able to permeate through the young sedimentary rock, and it was only later, during the process of diagenesis (the post-depositional physical and chemical changes in sedimentary rock), that hematite was formed in these ancient rocks. The existence of iron oxides is an important environmental indicator of Earth's evolutionary progress, and it provides evidence that substantial free oxygen had by this time become available in the atmosphere, eventually enabling organisms to start life on land.

Things to see and do

Drive the pass and look out for the geology: there are important geological features to discover – but note that straight vertical grooves were caused by drills during road construction. See the:

- **Abundant primary sedimentary structures:** These include trough cross-bedding, indicating that the deposition of sand took place on a flood plain crossed by multiple braided rivers.
- **Pebble beds and erratic pebbles:** This shows that shallow, high-energy depositional conditions existed on the river flood plain.
- **Concentrations of hematite:** Found along bedding planes and on cross-bedding, indicating that iron minerals were co-deposited within the sandy sediments.
- **Iron hydroxides:** Permeating the sandstone and moving along bedding planes and, at a later stage during diagenesis, converted to hematite.
- **Countless irregular, rounded or elongated bleached patches:** Also called 'reduction spots' where the rock has been locally leached. The alteration cross-cuts sedimentary structures and destroys diagenetic hematite, so it is clearly a much later reaction.

A thin, vein-like layer of black crystalline hematite has formed along the bedding surface, resulting from remobilisation of iron oxide from within the rock mass.

The cross-bedded, fluvial sandstone also shows lighter roundish patches that are called 'reduction spots'. These common but enigmatic patches are the result of the much later leaching and removal of iron by ground water.

GAUTENG AND NORTH WEST

OVERVIEW OF LANDSCAPES AND GEOLOGY

Covering much of Gauteng and part of adjoining North West, this region is of considerable geological age and richly endowed with mineral wealth. It includes the 'discovery area' of the Witwatersrand, and part of the platinum-rich Western Limb of the Bushveld Complex. Geomorphically it is made up of the Highveld Plateau in the south and the Bushveld Basin in the north, separated by the sinuous low mountains of the Magaliesberg. The focus is on Geosites 39–45, shown on the national maps (pp.12–13), and on the regional landscape and geological maps (pp.246–247).

The major geological formations of the region are listed below:

Basement Granite-Gneiss (Archaean)

The 'core' of the region is the oval-shaped inlier of ancient granites and gneisses, known as the Johannesburg Dome – the result of geological 'doming' and **Basement** uplift. About 45km across, it comprises several varieties of granodiorite in the south, and tonalitic gneiss and migmatite in the north, ranging from about 3,340–3,110 million years in age. This was the 'floor' on which the surrounding sedimentary formations were deposited, now seen as Witwatersrand rocks in the south, and Transvaal rocks along its flanks and in the north. Not shown on the map are several remnants of greenstone belts, close to the east-west-trending Witwatersrand ridge.

Witwatersrand Supergroup (Middle Archaean)

The Witwatersrand Basin started to form on continental crust around 2,950 million years ago, in response to the far-off collision between the Kaapvaal and Zimbabwe Cratons. Sedimentary deposition and reworking ended around 2,700 million years ago. The present basin is complex, roughly oval-shaped, some 350km long and up to 200km wide, and almost entirely covered by younger formations. Near its centre is the Vredefort Dome, scene of a massive meteorite impact, the effect of which was to enhance preservation of the already lithified rock formations of the Witwatersrand Basin (see pp.154–161) by forcing the strata deeper down into the crust over a wide area, so saving them from weathering and erosion. Geomorphically, the basin forms part of the elevated Highveld Plateau and the ridges of the Witwatersrand are a regional watershed, separating northwards drainage (ultimately towards the Indian Ocean) from southwesterly drainage (eventually into the Atlantic Ocean). The geology is best seen along its northern edge where the full 7,000m-thick, south-dipping sequence is exposed. It is divided into lower and upper divisions:

- **West Rand Group** (lower), which comprises roughly equal amounts of shale and quartzite with minor conglomerate layers, and can be seen at the Walter Sisulu National Botanical Garden (**Geosite 44**).
- **Central Rand Group** (upper), consisting mostly of quartzite plus thin but extensive gold-bearing conglomerate layers or 'reefs'. The discovery of the Main Reef in 1886 was in a sedimentary bed of pyritic quartz pebble conglomerate, close to Gold Reef City (**Geosite 45**).

■ Ventersdorp Supergroup
(Late Archaean)

Around 2,700 million years ago conditions changed on the Kaapvaal Craton and outpourings of Klipriviersberg flood basalt, the start of Ventersdorp continental volcanism, signalled the end of sedimentation of the Witwatersrand Basin. South and southeast of Johannesburg the hills are made of this lava, which makes an important part of the landscape of the Highveld Plateau. These rocks are also exposed around part of the Vredefort Dome (see pp.154–161).

■ Transvaal Supergroup
(Late Archaean to Early Proterozoic)

The extensive Transvaal Basin shows essentially the same groups and formations as seen in Mpumalanga (see Mpumalanga Overview, pp.162–163, for more on its formation). The following is relevant to geological sites in the region:

- **Black Reef Formation**, which forms a consistent marker horizon at the base of the **Transvaal Supergroup** and is present around the Johannesburg Dome, dipping outwards. The historic gold mine at Kromdraai, dating from 1881, is located within this formation, in the area of the Cradle of Humankind.
- **Malmani Subgroup** (lower part of the **Chuniespoort Group**), in which Malmani dolomite (or dolostone) almost entirely surrounds and dips away from the Johannesburg Dome, and continues westwards in a broad belt towards Ventersdorp, forming a gently undulating karstic land surface. The process is well illustrated at the Cradle of Humankind, where the cave-producing effect of karst formation can be seen at Sterkfontein Caves (**Geosite 43**) and at nearby Wonder Cave. The dolomite beds are of algal origin, the process of which is described on pp.163 and 184.
- **Pretoria Group** – a sequence, about 6,000m thick, consisting mainly of alternating quartzites and shales, and marking the start of a new period of Transvaal sedimentary deposition. It forms an east-west-trending, gently sloping, stepped (cuesta) landscape of north-dipping layers. Its age of around 2,220 million years is indicated by the volcanic **Hekpoort Formation** (andesite). Of main interest is the approximately 300m-thick **Magaliesberg Formation**, the resistant quartzite that makes the crest of the Magaliesberg range, as seen at Hartbeespoort Dam (**Geosite 40**), where intrusion of the hot Bushveld Complex magma caused metamorphic recrystallisation of the original sandstone. East of Pretoria the **Premier kimberlite** pipe (Cullinan Mine, **Geosite 42**) intrudes through quartzite and felsite of the uppermost **Rayton Formation**.

■ Bushveld Complex – Western Limb (Early Proterozoic)

A large part of the complex (see Mpumalanga Overview, pp.162–163, for more information) occurs directly north of the Magaliesberg range, where it is known as the mafic **Rustenburg Layered Suite** (RLS). It does not outcrop well but its subsurface presence is revealed by numerous chromite and platinum mines between Brits and Rustenburg, and beyond. Further north the presence of the upper part of the RLS and **Lebowa Granite Suite** is seen, quarried in places for decorative stone. The impressive meteorite crater at Tswaing (**Geosite 41**) occurs within pinkish, coarse-grained Bushveld granite.

■ Pilanesberg Complex
(Middle Proterozoic)

This large, ring-like intrusion, about 25km across, is located on the western side of the **Bushveld Complex**. Formed around 1,200 million years ago, it is one of the largest and best exposed intrusions of very uncommon alkaline rocks on Earth. The complex is located almost entirely within the bounds of the acclaimed Pilanesberg National Park, and there are many unusual rock sites that are signposted and described along some of the park's roads (**Geosite 39**).

Green foyaite beacon at geosite (**10**) *(see Figure 31, p.250) near the Bakubung Gate at the southern entrance to the Pilanesberg Park.*

Figure 29: Landscape map of part of Gauteng and North West. This clearly shows the different geomorphic regions: the Highveld Plateau in the south, in tones of brown; and the lower-elevation Bushveld Basin in yellow and green in the north. The two regions are separated by the sinuous Magaliesberg range.

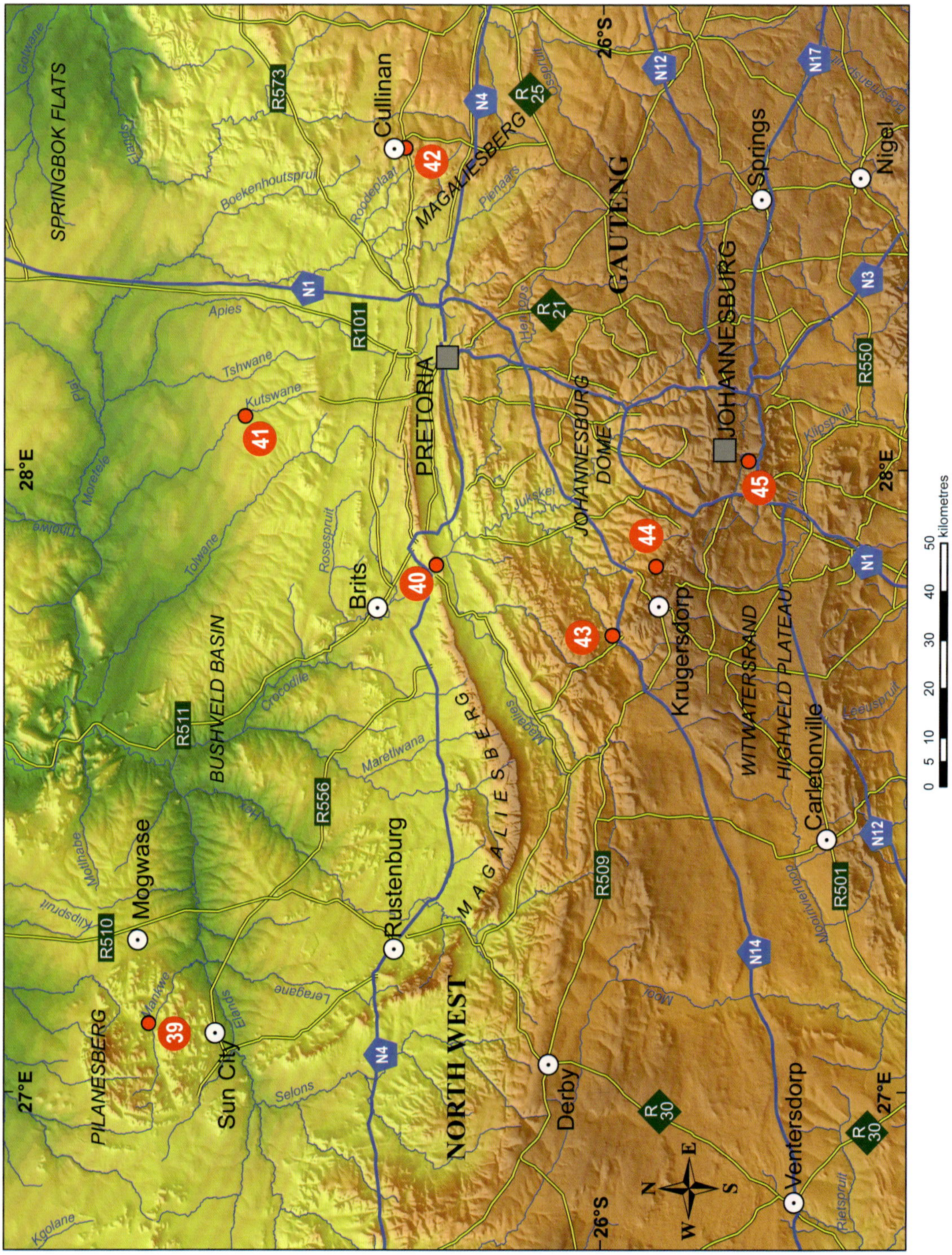

Figure 30: Geological map of part of Gauteng and North West. The different geomorphic regions are revealed by the underlying geology: the Highveld is dominated by the Witwatersrand and Transvaal Supergroups, while the Bushveld Basin is largely underlain by rocks of the Bushveld Complex, the circular massif of the Pilanesberg to the northwest, and Karoo geology in the Springbok Flats to the northeast.

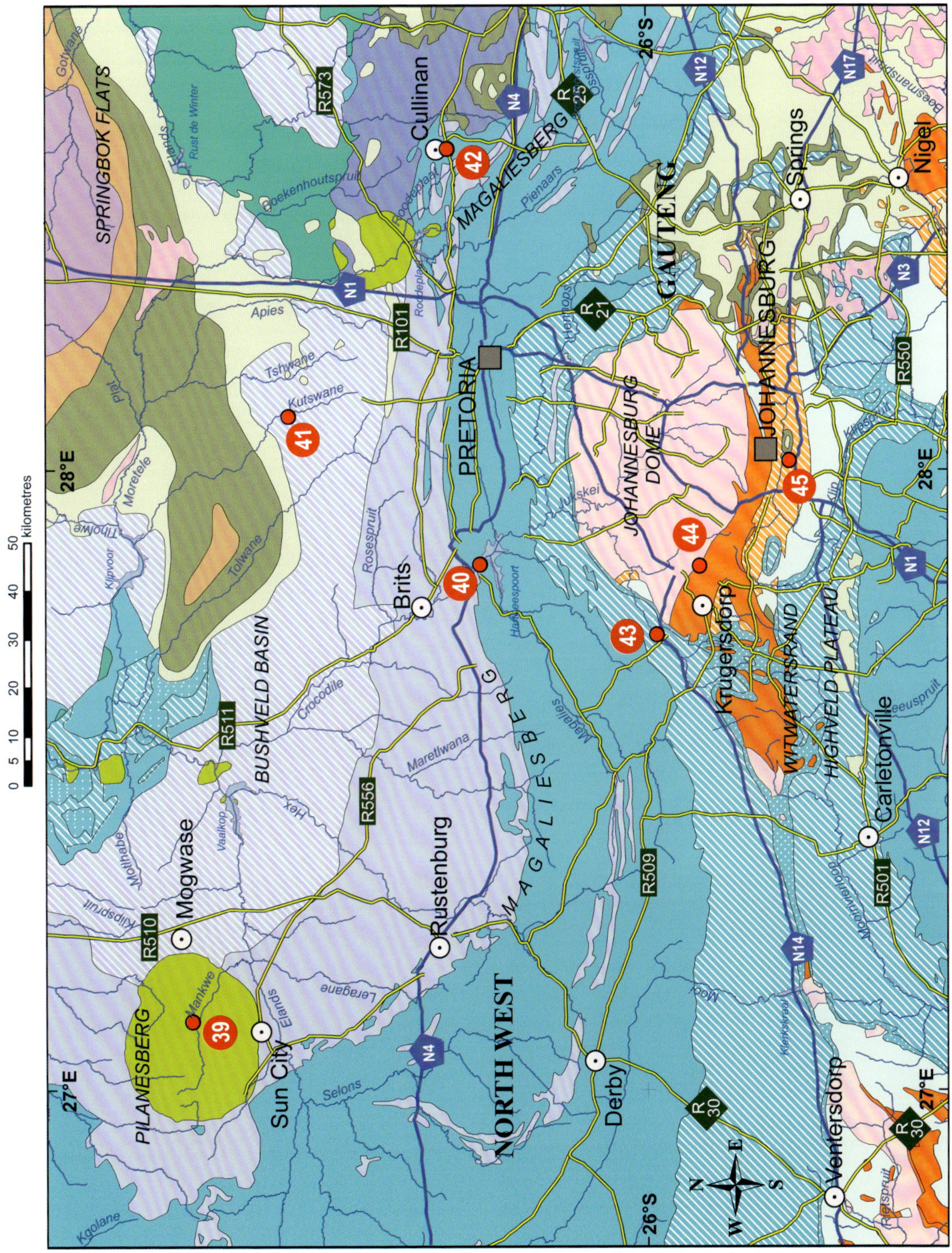

• A stratigraphic key to this simplified regional geological map appears on the inside covers

39 Pilanesberg Complex

NORTH WEST PROVINCE NATIONAL PARK

The best exposed and third-largest alkaline ring complex in the world

Located more or less in the centre of the Pilanesberg Complex, man-made Mankwe Dam is the lowest point, and is surrounded by a roughly ring-like arrangement of hills and valleys of various alkaline igneous rocks and lavas.

In 1977 the Tswana-speaking Bakgatla people agreed to the establishment of the Pilanesberg as a unique conservation area. The Pilanesberg was vacated, environmentally rehabilitated and redeveloped into a game reserve, and declared the Pilanesberg National Park in 1984. 'Operation Genesis' took place in the early 1980s, a very ambitious wildlife relocation programme during which over 6,000 indigenous animals were re-introduced, restoring most of the animal biodiversity to the area. The mountains are named after Kgosi Pilane, an important 19th century leader of the local Bakgatla people.

■ Location and getting there

■ The Pilanesberg massif stands high above the rather featureless Bushveld plain, starting about 40km north of Rustenburg. There are several routes to get into the park: from the south use the R565 and enter at Bakubung Gate; from the east use the R510 and enter at Manyane Gate; and in the north there is Bakgatla Gate near Moruleng (formerly Saulspoort). The park is managed by the provincial North West Parks and Tourism Board.

■ GPS (Pilanesberg Centre): 25°14′55″S; 27°06′35″E

Setting the scene

The Pilanesberg massif is almost circular in outline – this is superbly illustrated on the Landsat image (p.251). It measures about 28km from east to west and 24km from north to south, and covers about 530km^2. Over millions of years nature has sculpted it into a series of ring-like mountains that provide a natural sanctuary of diverse habitats, which are home to a wide range of animal life. Described by geologists as the Pilanesberg Alkaline Complex, it is the largest of this type of igneous intrusion in Africa, and the third-largest in the world. It is generally accepted that it represents the remains of an enormous volcano, probably larger in area and higher than Mount Kilimanjaro. Although similar large, ring-like alkaline intrusions are known from elsewhere, such as the Kola Peninsula in Russia, no other place in the world provides such a rewarding self-drive geological tour through an alkaline ring complex.

Regional geology

The **Pilanesberg Complex** was intruded through the much older rocks of the **Western Limb** of the **Bushveld Complex** around 1,200 million years ago. There are two main components to the geology: an older series of volcanic rocks that built the original volcano, and the deeper root zone where related alkaline magma was intruded far below the surface. Most impressive are the hills that comprise a very unusual suite of coarse-grained plutonic rocks, although the enigmatic contacts between the formations are not visible due to more recent cover. The dip of the formations, best seen on aerial photographs, is generally inward but steeply outwards in the north. A major fault bisects the complex.

Landscapes and landforms

The Pilanesberg stands out because the rock formations are relatively resistant to weathering and erosion. In spite of this, over many millions

Figure 31: Simplified geological representation of the Pilanesberg Complex and geosite location.

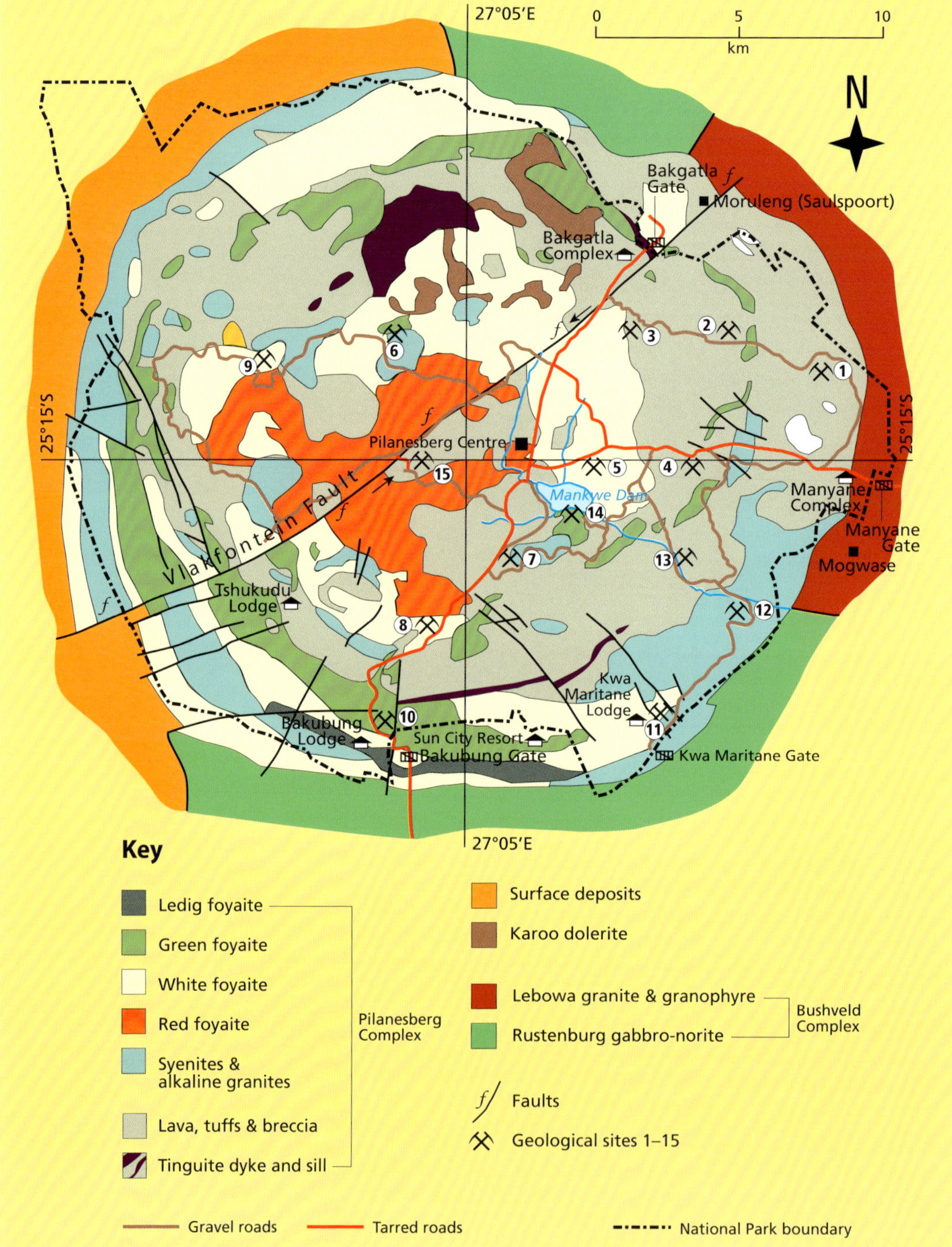

Composite map adapted from *The Geology of South Africa*, Council for Geoscience (2006) and *Pilanesberg*, Jacana (1997)

A simplified geological map of the Pilanesberg Complex shows the roads and location of 15 demarcated geological sites that can be visited. Other than the roughly ring-like arrangement of steeply dipping geological formations, the dominant feature is the major northeast-southwest-trending fault that neatly bisects the complex.

Michael Cronwright, The MSA Group

This false-colour Landsat 5/Aster DTM image clearly shows the diverse and rugged terrain of the Pilanesberg Complex (about 24km from north to south), and should be compared with the simplified geological map opposite.

How was the Pilanesberg Complex formed?

This is a controversial topic and not all geologists are agreed on the mechanism of its formation. The conventional explanation is that the Pilanesberg Complex is largely volcanic, formed by a series of eruptions around 1,200 million years ago that produced lava flows, agglomerates and tuff beds that ultimately collapsed inwards. At depth, and following a number of large-scale, ring-like fractures, pulses of alkaline magma forced their way upwards to form several thick, funnel-shaped foyaite intrusions and plugs of syenite and alkali granite. This theory poses a major problem: how space for these very large ring-like intrusions was created; and the answer to this is still unresolved. At some early stage, caldera-like subsidence of the volcano took place along conceptual ring faults, followed by radial fracturing that formed a pattern of sub-parallel dykes, which occur mostly to the northwest and southeast of the complex.

Not everyone accepts the funnel-shaped intrusion explanation and an alternative view sees the Pilanesberg's foyaites and syenites as part of a thick and more gently dipping, saucer-shaped lopolith – a large and deep igneous intrusion that's lenticular in shape and that later subsided in the centre. Indeed, this is still a fertile field for geological research.

of years they have been reduced to a remnant of their previous extent, and now form a series of large, ring-like – if not always continuous – rocky hills and mountains. Towards the northern side a peak known as Matlhare is the highest point, rising to 1,687m; and, at 1,120m, the man-made Mankwe Dam lies in the central plain of the park. Drainage is radially outwards and numerous faults have created valleys, notably the major Vlakfontein Fault, which effectively bisects the complex, extending from the southwestern margin northeast towards Moruleng, creating the Renosterspruit Valley.

Rock formations

- The ring-like hills are the result of at least five funnel-shaped intrusions, stacked one on top of another, and made of varieties of a rare igneous rock called 'foyaite'.
- Foyaite is a type of coarse-grained syenite containing alkaline feldspars, nepheline (a rare sodium aluminium silicate) and sodium-rich pyroxenes and amphiboles, and no free quartz.
- The foyaites differ in their mineral proportions and colour but have similar chemical compositions. They are called 'alkaline' as they are composed largely of potassium- and sodium-rich silicate minerals and are relatively poor in silica.
- From the centre outwards, different intrusions are recognisable (see geological map on p.250, **Figure 31**, for distribution and formal names) but the 'rings' are not complete.
- There are also intrusions of nepheline syenite, plug-like bodies of syenite in the core, a very small plug of alkali granite, and even a small kimberlite.
- Considerable volcanic activity took place during the early eruptive phase and is represented by a series of fine-grained, dark and light-coloured outpourings of phonolitic and trachytic lavas, breccias, agglomerates and stratified tuffs, the latter being fall-out from explosive volcanic eruptions.
- A number of the formations are unusually rich in fluorite, thorium, strontium, zirconium and uranium, and rare earth elements.

*An impressive red syenite intrusion, as seen at geosite (**7**) (**Figure 31**), occurs near the centre of the Pilanesberg. Trial quarrying was carried out here for production of an attractive dimension stone, but not taken further as it contained small grains of iron oxide. It later became part of a game reserve and it is now an important and very popular national park.*

A specimen of layered uranium-bearing tuff displayed at geosite (**13**) (**Figure 31**). It contains rock fragments, indicating its origin from volcanic ash and debris. It also contains low levels of uranium and rare earth minerals, and is mildly radioactive.

Outcrops of coarse-grained white foyaite are common around the Manyane area. The matted texture of white feldspar laths, grey nepheline and dark pyroxene minerals indicates slow crystallisation of the magma.

This attractive specimen of polished red syenite comes from geosite (**12**) (**Figure 31**) *in the southeast, close to KwaMaritane. The rock polishes well but has not been commercially exploited.*

Things to see and do

- **There are 15 demarcated geological sites in the Pilanesberg** (marked 1–15, see map, p.250): At each site a polished specimen is mounted on a stone column together with a descriptive plaque. The sites are easy to locate but a drawback is that you are not allowed to get out of your vehicle to look more closely at the rocks unless accompanied by an armed ranger. Most of the plaques (provided by FOPS – Friends of the Pilanesberg Society) are in reasonable condition but some are missing. To get a better sense of the underlying geology you should also refer to the comprehensive general guidebook that is available at the entrance gates.

The rock pedestal at geosite (**3**) (**Figure 31**) *in the northeastern part of the complex; the polished rock specimen is of dark, slightly porphyritic, fine-grained lava, from the volcanic phase of the complex.*

40 Hartbeespoort Dam

PART OF PROPOSED BIOSPHERE RESERVE

Ancient mountains of hard quartzite, a 'cuesta' landscape and a fault-made river gorge

The Magaliesberg range, as seen looking west from the upper cableway station, with part of Hartbeespoort Dam in the distance. Gently north-dipping Magaliesberg quartzite, overlying Silverton shale, forms a 'cuesta' landscape. This prominent range effectively separates the Highveld Plateau in the south (left) from the Bushveld Basin to the north (right).

The Hartbeespoort is a narrow gorge where the Crocodile River has cut its way through the imposing Magaliesberg with the help of a geological fault – one of the few rivers that has succeeded in doing this. It is in this gorge that, in the 1920s, a major dam was built for irrigation purposes. The gorge is also the gateway to the Bushveld Complex and the semi-tropical bushveld environs.

Location and getting there

- Hartbeespoort Dam is located in the east-west-trending Magaliesberg range, about 35km west of Pretoria and some 65km northwest of Johannesburg. Various routes get you to the dam wall: the easiest takes you via Schoemansville on the R511, then follow the winding road to the dam wall and tunnel.
- GPS (parking area west of dam wall): 25°43′32″S; 27°50′49″E

Setting the scene

The Magaliesberg forms a mountainous ridge running from northwest of Rustenburg to east of Pretoria, more than 200km long, and is a major feature of the Highveld landscape.

In 1913 an ambitious plan was developed to dam the narrow gorge on the Crocodile River south of Brits, to provide for downstream irrigation, but it was only in 1923 that the dam was completed. Buttressed on each side by massive outcrops of hard quartzite, the curved

The fault-created gorge (the original Hartbeespoort) on the Crocodile River is now part of Hartbeespoort Dam, here seen from the dam wall, looking south.

The dam wall, completed in 1923, is straddled by a decorative arch; built in the classical 'triumphal' style, it is adorned with Latin inscriptions.

The Bushveld Complex

North of the Magaliesberg is the Western Limb of the Bushveld Complex, part of the world's largest layered intrusion, and one of the choicest bits of geological real estate in the world – here lie the world's richest deposits of platinum metals, chromium and vanadium. After the sedimentary deposits of the Transvaal Basin had become sandstone and shale, an underlying, deep-seated mantle hot-spot developed, and the overlying part of the Earth's crust began to weaken. Around 2,060 million years ago, a period of unprecedented igneous activity started as thin intrusions of mafic magma penetrated the sedimentary layers forming the diabase sills we see today. This was soon followed by the injection of enormous volumes of magma, deep into the central portion of the Transvaal Basin, forming the mafic phase Bushveld Complex. Slightly later, Bushveld granite was also intruded. The weakened crust and the vast mass and heat of these intrusions resulted in structural sagging of the entire Transvaal Basin, forming a consistent inward dip. The great heat caused the neighbouring sedimentary rocks to become thermally metamorphosed to quartzite and hornfels.

When viewed facing northwards down the Crocodile River from the dam wall, the river can be seen to follow the fault that created the gorge.

The Crocodile River downstream below the dam wall, which can be seen in the far distance. Kosmos Mountain is on the right.

dam wall is 101m long and 60m high. With a shoreline of around 56km, Hartbeespoort Dam is the largest body of water in the Gauteng region. Desirable urban settlements have developed in the surrounding areas and the dam has become a favourite recreation venue for the land-locked population. It is locally known as the Bushveld Riviera.

Regional geology

The Magaliesberg is made of the upper formations of the mainly sedimentary **Pretoria Group**, part of the **Transvaal Supergroup**.

On the western side of the original Hartbeespoort is Kosmos Mountain, which shows a more easily weathered diabase sill in between beds of Magaliesberg quartzite dipping north at 25°.

At Hartbeespoort Dam these formations dip consistently at around 25° to the north. The thick quartzite unit that forms the top of the mountain ridge is the **Magaliesberg Formation**, and beneath this, poorly outcropping shale of the **Silverton Formation** forms the lower southern slopes and the broad Moot Valley. The original 200m-wide Crocodile River gorge follows a north-northwest-trending fault, which is part of a regional down-faulted structure called the Brits Graben. The other parallel fault of the graben lies some 3km to the west at Commando Nek. Between these two faults the upper crust was depressed by several hundred metres, thus displacing to the south a short section of the Magaliesberg ridge. Over millennia, considerable erosion followed and what remains of the graben is now prominently exposed as Kosmos Mountain with the Hartbeespoort fault on its eastern side.

Landscapes and landforms

The Magaliesberg is an eroded remnant of the surface that existed before the African land surface was formed. Today, Hartbeespoort Dam is overlooked by the Magaliesberg ridge, which reaches 1,600m. Part of the landscape can be traced back to the time of Dwyka glaciation around 300 million years ago, when ice sheets partly planed down the mountains and carved wide glacial valleys. The range is a fine example of 'cuesta' landscape (also called 'bankenveld'), an asymmetric landscape with a steep scarp-slope and a gentle dip-slope.

Rock formations

- Coarse-grained Magaliesberg quartzite originates from sandstone that was deposited less than 2,200 million years ago as clean

North-dipping beds of Magaliesberg quartzite at the upper cableway station. Inset: Showing that the quartzite was coarsely recrystallised (thermally metamorphosed) by the enormous heat generated from the Bushveld Complex which lies nearby, to the north.

quartz sand along a shallow shoreline, with occasional sandy flats on which ripple marks formed and, in places, are now preserved.

- Thermal metamorphism from the nearby Bushveld Complex resulted in the original sandstone being completely recrystallised to form large interlocking quartz grains and coarse-grained quartzite.
- The Silverton shales were deposited as mud in deeper water within the vast inland Transvaal Sea, and much later several sills of Bushveld-related diabase were intruded into the shale.

Ripple marks indicate that this sandstone was deposited in very shallow water, and has been reformed into hard, recrystallised quartzite with ripples still preserved.

The entrance to the single-lane road tunnel that was blasted through very hard, recrystallised Magaliesberg quartzite, on the eastern side of the dam wall.

Things to see and do

- **Do the aerial cableway:** This is the highlight of any visit. In 2012 the new aerial cableway was opened and it takes you to the top of the Magaliesberg, at a height of about 1,600m. From there you will have a superb view across Hartbeespoort Dam, and a panoramic view of the entire region, from the Witwatersrand in the south to the Pilanesberg in the northwest.
- **Visit the dam wall and tunnel:** Walk across the dam wall and see the 57m-long single-lane road tunnel blasted through massive quartzite. Notice the north-dipping, very hard, coarse-grained quartzite, and look down the fault-created Crocodile River gorge (Hartbeespoort). Parking is available on the northwestern side.

An attractive landscaped indigenous garden has been established at the upper cableway station.

The new aerial cableway and the viewing deck, which provide superb views to the south of the Magaliesberg, including Hartbeespoort Dam and Schoemansville.

41 Tswaing Meteorite Crater

AN EXTRATERRESTRIAL LANDING SITE

One of the world's best preserved and most accessible small, bowl-shaped impact sites

Terence McCarthy

An aerial view of the Tswaing Meteorite Crater from the northern side, with its surrounding preserved bushveld vegetation. Previously known as the Pretoria Salt Pan, this enigmatic landmark was at one time thought to be a gaseous volcanic crater.

Previously known as Zoutpan or the Pretoria Salt Pan, the crater's modern name means 'place of salt' in Setswana. For many years geologists thought that it was a brine-filled vent related to a gaseous volcanic explosion but, after many years of scientific controversy, deep drilling within the crater in 1989 at last confirmed its extraterrestrial origin.

Location and getting there

- Tswaing Crater is located some 40km north-northwest of Pretoria, near the major settlement of Winterveld, and within a 1,946ha conservation area managed by Ditsong Museums. From Wonderboom the suggested route is via the M1, then the M35; drive past Onderstepoort and through Soshanguwe until you reach the turn-off marked 'Tswaing'.

- GPS (SE viewpoint): 25°24'38"S; 28°05'14"E

The crater, about 1,100m across, seen from the popular viewsite on the southeastern side, close to the old wagon road into the crater. The small sand spit at water level on the far side is the site of the 1989 borehole. Notice the surrounding elevated and forested rim, made of upturned granite and ejected granite boulders.

Roger Price

Setting the scene

About 220,000 years ago an unwelcome visitor from space arrived on Earth, resulting in a cataclysmic event. Early humans living in the region would, for just a few seconds, have seen an enormously bright light across the sky, then heard an explosion of unparalleled force and felt the ground quake beneath their feet. A massive shock wave would have wiped out all life in the surrounding area and an immense dust cloud would have arisen, blotting out the Sun for many days. Earth had taken a rare hit by a meteorite of the stony type, not more than 50m in size, but travelling at hyperspeed – perhaps at more than 30km per second.

Regional geology

The surrounding rocks are **Nebo granite** of the 2,060-million-year-old **Bushveld Complex**. Core drilling within the crater took place in the late 1980s: the borehole first passed through 90m of soft saline mud and carbonate sediments; it then penetrated unconsolidated

Figure 32: Schematic cross-section through the Tswaing Crater.

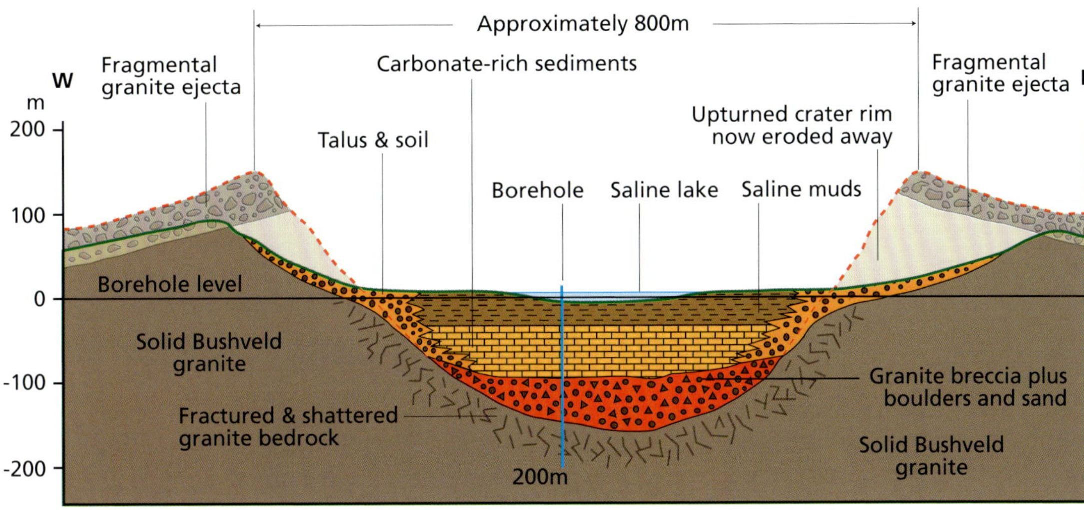

Adapted from *The Geology of South Africa*, Council for Geoscience

A schematic east-west cross-section through the bowl-shaped crater, based on evidence from surface mapping and a borehole drilled in 1989. The original crater rim has been lowered by erosion (to today's surface, the green line) and the crater partly filled by granitic debris, fine-grained, saline-rich muds and carbonate sediment.

Salt and soda extraction

Tswaing has been an important source of salt for animals and humans for thousands of years. The salts within the crater sediments came from leaching of the surrounding granite by ground water that collected in a closed basin and became concentrated as a result of evaporation. In addition to common salt, the main saline residues, produced over thousands of years, are the natural sodium and calcium carbonate minerals called 'trona' and 'gaylussite'. Commercial extraction started in 1912 and continued to 1956, with numerous ups and downs in production. The soda-salt brine was pumped to a plant on the surface where, by means of a complex and innovative process, both soda ash and common salt were extracted.

sandy material with granite fragments, followed by fractured granite and, at a depth of 200m, went into coarse-grained pinkish Nebo granite (see cross-section, **Figure 32**). These results dispelled all ideas that the crater was of volcanic origin.

Landscape and landforms

The crater rim is now about 1,100m across and is well covered by trees and bushes. It rises gently to around 60m above the surrounding sandy granitic plain but the inner part of the rim is much steeper and drops some 120m into a shallow, brine-filled crater lake underlain by salt-rich, clayey sediment.

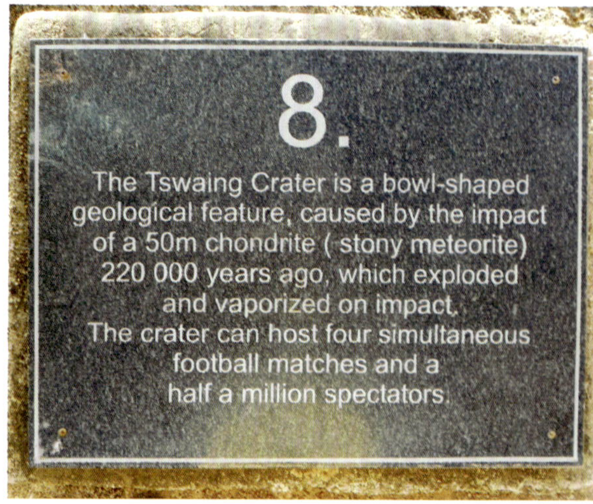

A rock plaque (No. 8 on the crater trail) located at the best viewsite on the southeastern side briefly explains the origin of the crater.

Christian Koeberl

Wolf Uwe Reimold

There have been several other meteorite impacts in southern Africa. These images show the circular 2.5km-wide Roter Kamm crater in southern Namibia, 3.4 million years old and largely covered by desert sand. Inset: A false-colour Landsat image of the crater. This location is very difficult to reach.

Rock formations

There are two very different aspects to the geology: the original Nebo (Bushveld) granite and its fragmentary debris; and the salt-rich sedimentary deposits that later formed within the crater.

- Granite fragments, large and small, represent rocks that were ejected at the time of explosive impact. The presence of microscopic deformation features in quartz grains and minute spherulitic glassy droplets formed by impact melting confirm the meteorite impact origin. Deeper down there is sand with small granite fragments followed by granitic breccia.
- Younger finely layered clays made of salt- and carbonate-rich muds were derived from the accumulation of fine sediment and salts washed into the closed crater-like system. They contain micro-fossils that hold important information on the palaeo-environment over the past 200,000 years.

On the southeastern side of the Tswaing crater rim are randomly oriented granite blocks that were ejected from the crater at the time of the explosive meteorite impact. This is the most popular viewsite of the crater.

Tswaing crater formation

Meteors are small bodies of planetary debris that are constantly moving about in space and show up as shooting stars when they burn up in the Earth's atmosphere. If they reach Earth's surface they are known as 'meteorites', and Tswaing is a world-class example of an impact site. The immense kinetic energy released at the time of impact explosively excavated a crater about 800m across and 200m deep. We cannot fully appreciate the devastation that was caused, certainly covering more than 1,000km^2. Within this area, the catastrophe would have obliterated all life. This is what would have happened on impact:

- Granite bedrock was highly compressed, intensely heated and fractured, and the stony meteorite was completely vaporised.
- An explosive reaction followed in which granite blocks were ejected, leaving a large, bowl-shaped crater with an upturned granite rim.
- Blocks of granite and mineral fragments fell over the surrounding area, forming a thick 'ejecta' blanket outside of the crater, and debris also fell back into the crater.
- Rain and ground water collected in the crater, slowly vegetation took over, and erosion and lowering of the rim into the crater produced the salty and carbonate mud deposits.

Things to see and do

- **Visitors Centre:** The crater has a long history of human usage dating back to the Middle Stone Age. There are plans to develop the site into a major tourist attraction, along with an educational facility.
- **Tswaing walking trail:** The 7km-long trail is the highlight of any visit. This follows the eastern crater rim, continues down to the crater floor, and returns via the steep old wagon road, with interesting geology along the way.
- **Nature reserve:** The indigenous vegetation on the farm is little changed since the early days, making this an ideal habitat for wildlife, and there is an enormous diversity of bushveld trees, along with excellent birding.

Fragmented and rearranged weathered granite blocks along the old wagon road into the crater represent solid granite that was broken up during the impact event.

A close-up view of the original floor rocks: coarse Bushveld (Nebo) granite, consisting mostly of glassy quartz and pink feldspar crystals, now partly shattered. This is a fresher example of the rocks seen along the old wagon road.

Roger Price

The 1989 drill-site is at the end of the small sand spit in the saline lake. Material recovered from this borehole presented critical evidence about the crater's meteoritic origin.

Part of the old wagon road into the crater, along which ox-wagons originally transported their salty load out of the crater.

42 Cullinan Diamond Mine

FAMOUS FOR THE WORLD'S LARGEST DIAMOND

South Africa's largest kimberlite pipe and source of many of the world's biggest diamonds

De Beers

The old open-pit mine seen from the southern side, with the modern-day mining plant and headgear in the distance. The sidewalls are slowly collapsing into the hole, which was mined to a depth of 189m. The scale of the site is deceptive – the hole itself is about 1km across.

Previously known worldwide as the Premier Mine, this mine is renowned for having produced the fist-size Cullinan Diamond in 1905, the largest gem diamond ever found, weighing 3,106 carats (0.62kg). The stone was cut into nine major and 96 minor gems; the four largest are part of the British Crown Jewels. The mine has yielded many other large diamonds, including the world's biggest cut diamond, the 'Golden Jubilee' of 545 carats. Now an underground mine, it continues to produce amazing diamonds and its riches are expected to last well into the future.

Location and getting there

- The mine lies within the village of Cullinan, some 30km east-northeast of central Pretoria. To get there from Pretoria either take the N4 going east, then turn off after about 30km onto the R515 to Rayton; or take the N1 north and at the Zambezi Drive intersection go east on the R513 for about another 30km.
- **GPS** (car park off Oak Street): 25°40′15″S; 28°31′01″E

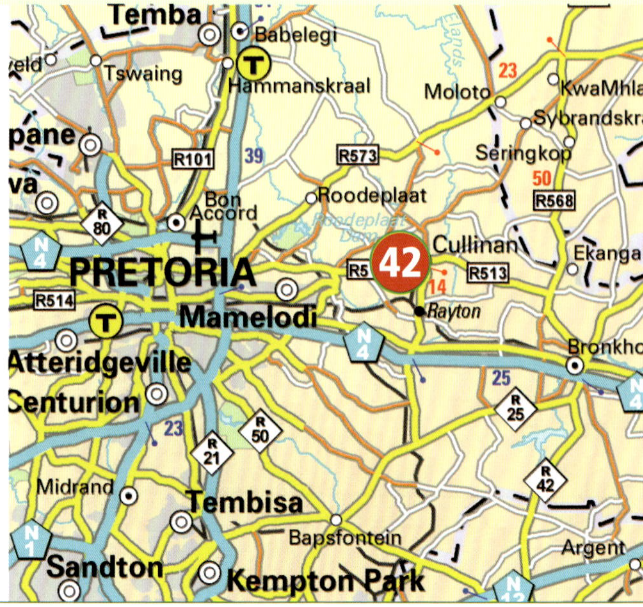

Setting the scene

Picture this scene: it is 1898 and two men, one an elderly digger named Percival White Tracey, and the other Thomas Major Cullinan, a wealthy businessman and amateur prospector, are standing on a farm about 33km east of Pretoria. One holds a beautiful blue-white, 3-carat diamond recently found nearby. Given the absence of kimberlite on this farm, it's clear that the diamond must be an alluvial stone washed down a mostly dry stream that wends its way through a neighbour's farm, Elandsfontein, on which the difficult owner, Willem Prinsloo, has strictly forbidden all prospecting. However, lured by the excitement of unknown riches, Cullinan starts negotiations to buy what he senses to be a potential source of diamonds. Negotiations about the farm are interrupted by the Anglo-Boer War, which rages from 1898 to 1902, but after this, Cullinan persuades

From the collection of John Lincoln at Cullinan, photographer unknown (1902)

A prospecting panning operation on the Premier kimberlite around 1902, before mining started. On the left is Fred Wells who, in 1905, recovered the famous Cullinan Diamond.

De Beers

Premier Mine dignitaries in 1903. Front, left to right: Fred Wells, Thomas Cullinan, Percival Tracey and unknown; back, left to right: William McHardy and unknown.

Diamond mining since 1903

The pipe's pre-mining surface area was 32ha and Cullinan's resulting Big Hole is now nearly 1,000m long and 450m wide – at least four times the size of the Big Hole at Kimberley – and it continues to grow as the sidewalls collapse into the old open-pit. Since mining started from the surface in 1903 more than 350 million tons of kimberlite have been mined to produce around 120 million carats (more than 24 tons) of diamonds, and mining continues to this day. Open-pit mining ceased in 1932, having reached a maximum depth of 189m, and in 1947 a start was made on the first shafts for underground mining, now done using large-scale mechanised block-caving methods.

Maggie Newman / De Beers

Artist's impression of the Premier kimberlite at the time of the last major intrusion some 1,200 million years ago. A composite volcanic pipe with at least four different types of kimberlite was formed.

Maggie Newman / De Beers

A cut-away impression through the kimberlite pipe today showing geology and the underground block-caving mining method. The floor of the open-pit will continue to be lowered as underground mining progresses.

Some famous diamonds

Of all kimberlite pipes worldwide, the Premier kimberlite pipe has produced more than 25% of diamonds over 400ct. Some examples are:

De Beers

Golden Jubilee: rough 755ct, cut 545ct

Replica of the Cullinan: rough 3,106ct, plus other gems

De Beers

Premier Rose: rough 354ct, cut 137ct

De Beers

De Beers

Centenary: cut 273ct (above left), rough 599ct, cut 273ct (right above and below)

An archive photo of the early No. 1 working at the Premier Mine, probably around the time that the Cullinan Diamond was dug out by Fred Wells.

A historic photograph, taken in 1905, of Thomas Cullinan, William McHardy (General Manager, holding the 3,106ct Cullinan Diamond) and Fred Wells, the finder. Notice the size of the Cullinan stone!

An aerial view of Cullinan Mine today showing the old open-pit, the mine headgear and the extensive recovery plant. The mine has been operating since 1903 and will still yield diamonds for many years to come.

the Prinsloo family to sell him their farm for £50,000. In 1902 Thomas Cullinan establishes the Premier (Transvaal) Diamond Mining Company. Wasting no time, the company soon starts diamond production, and in 1903 over 100,000 carats are recovered – an output of diamonds that soon rivals that of De Beer's combined Kimberley mines.

Regional geology

Cullinan's kimberlite pipe, named Premier, is 1,200 million years old, an unusually ancient kimberlite – in fact the oldest diamond-producing kimberlite in the world. At surface it intrudes through a reddish felsite layer that may be correlated with the 2,060-million-year-old **Rooiberg Group**, part of the Bushveld magmatic event. Originally covering 32ha (about 64 rugby fields), it is the largest of a cluster of 11 kimberlite pipes found in the Cullinan-Rayton area, and it is also South Africa's largest kimberlite pipe – the carrot-shaped intrusion formed by a gaseous explosion as magma rises very rapidly to the surface (see left-hand image in 'Diamond mining since 1903', p.267).

Rock formations

- The kimberlite pipe is roughly kidney-shaped in plan, and it is a multiple intrusion: it contains several varieties of kimberlite, all diamond-bearing, including the Grey, Brown and Black kimberlites.

- Large masses and fragments of Waterberg quartzite, a once overlying formation, were found within the pipe, which was also cut through by a 75m-thick gabbro sill.
- Typical of kimberlite pipes, it narrows progressively downwards and, at 1,000m below the surface, it splits into two root-like intrusions.
- It was formed by violent eruptions (see illustrations on p.267) as water- and gas-rich ultramafic magma from the Earth's upper mantle moved rapidly upwards through cracks in the crust and, when near to the surface, erupted explosively.
- This took place several times, producing the several different types of kimberlite with varied diamond content.
- Since the final eruption, at least 300m of kimberlite has been removed by erosion, together with the surrounding rock formations, no doubt liberating countless diamonds across the landscape.

Petra Diamonds

Another aerial view of Cullinan Mine looking northwards from the west side of the old open mine, and the extensive diamond recovery plant.

Things to see and do

- **Do the surface mine tour:** No visit to Cullinan is complete without doing this tour, which is conducted by knowledgeable local guides. It includes an explanatory video on mining and diamonds, seeing the collection of replica diamonds and photos, and viewing the vast open Cullinan Big Hole.
- **See diamond cutting:** You can do a walk-through visit of the on-site diamond cutting works and the jewellery shop for that ultimate 'diamonds are forever' purchase.
- **Visit McHardy House Museum:** This fully furnished, authentic Edwardian house is the original General Manager's residence. There are many other historic stone and corrugated iron buildings in the village dating from the early 1900s and now converted to quaint shops and restaurants.

Cullinan Mine headgear today. The mine is now owned and operated by Petra Diamonds.

The old stone-built miners' cottages along Oak Street, now used as shops and restaurants in what is a major tourist attraction within the Dinokeng tourism area.

43 Sterkfontein Caves

WORLD HERITAGE SITE

The richest single apeman fossil site in the world
and one of a cluster of many dolomitic fossil-rich caves

The surface limestone workings and archaeological walkways at Sterkfontein Caves where the almost perfectly preserved hominid (apeman) skull, which was soon nick-named 'Mrs Ples', was blasted out of the cave breccia wall by Robert Broom in 1947. Since then, many important hominid fossils have been found, including 'Little Foot' in 1997.

The fossil-rich dolomite caves around Sterkfontein are the most prolific source of apeman remains in the world, resulting in this unique heritage area being named the Cradle of Humankind. Sterkfontein Cave is the best known, having played a fundamental role in putting Africa at the centre of human evolution.

Location and getting there

- Sterkfontein Caves, owned by the University of the Witwatersrand, are located within the small Isaac Edwin Stegman Nature Reserve about 10km north of Krugersdorp and some 40km northwest of Johannesburg. The easiest route to get there is from the M47/N14 (Pinehaven) intersection near Muldersdrift. Then follow the N14 for 11km to Oaktree, take the R563 to Hekpoort and turn right after 1.5km; the well signposted site is a short distance further on and is open daily.
- GPS (caves entrance): 26°00′59″S; 27°44′03″E

Setting the scene

The Sterkfontein Caves were discovered in 1894 by lime-workers looking for deposits of pure limestone (travertine), which occurs as spectacular cave formations. During digging the lime-workers discarded deposits of hard cave breccia that contained countless fossilised bones. These fossils are of such extraordinary scientific value that in 1999 the surrounding 470km^2 area was declared South Africa's first World Heritage Site. It contains 15 excavated fossil-bearing cave sites, including sites such as Swartkrans, Kromdraai, Gladysvale, Drimolen, Coopers, Plover's Lake, Motsetse and Malapa, none of which is generally open to the public.

A view to the west from Sterkfontein Cave across to another archaeological dig site on the hillside in the distance. This is Swartkrans Cave, the second-richest hominid fossil site in the area, and also famous for evidence of the controlled use of fire.

Regional geology

All the caves are located within dolomitic limestone (or dolostone) of the **Malmani Subgroup** of the **Transvaal Supergroup**, outcropping on the northwestern side of the geological feature known as the Johannesburg Dome. The extensive Sterkfontien cave system overlaps the contact between the **Monte Christo** and **Oaktree Formations** in the lower part of the Malmani dolomite sequence, which was deposited around 2,500 million years ago within the shallow sea that filled the Transvaal Basin. As a result of later uplift of the Johannesburg Dome the dolostone beds dip from 25–30° to the northwest.

The complex cave system was formed (by a process known as 'karsting') by ground water slowly dissolving the dolostone along westerly- and northerly-trending joints or faults (see text box, p.84). Other than stromatolites of algal origin, these ancient rocks themselves contain no fossils.

Landscape and landforms

Around 3–2 million years ago this landscape consisted of tropical gallery forest fringed by savanna woodland, an ideal habitat for families of the apeman and other primates; but today it is part of the temperate grassy Highveld Plateau. This undulating rocky terrain lies at an altitude of around 1,500m and is underlain by dolomite layers, either chert-rich or chert-poor varieties. Numerous caves have developed beneath the karstic land surface and entries can often be recognised on the surface by increased bushy vegetation. Sterkfontein Cave lies in the valley of the Blaauwbank Spruit, a minor northeast-flowing tributary of the Crocodile River, and the main entrance is on a small hill at 1,491m. The relatively subdued surface expression of the cave belies its considerable underground extent – to a depth of about 50m and some 250m east-west – and its complexity. Nearby and slightly higher in the sequence is the smaller Lincoln Fault cave

Looking eastwards from Sterkfontein towards the valley of the Blaauwbank Spruit across the road on the left. It is hard to believe that below this land surface of outcropping Malmani dolomite lies the large and complex Sterkfontein Cave system.

The apeman fossils

The apeman (also called a 'hominid') is known only from its fossil remains. These primates were upright-walking, small-brained, very distant relatives of humans that in southern Africa developed along different evolutionary lines, starting around 3 million years ago. The first adult hominid fossil specimen was found at Sterkfontein in 1936 by the remarkable Dr Robert Broom, and in 1938 a different hominid species was found at nearby Kromdraai Cave. But it was Broom's discovery at Sterkfontein in 1947 of 'Mrs Ples' that created so much excitement; it was the almost perfectly preserved skull of an adult hominid to which he gave the name *Plesianthropus transvaalensis* (hence the popular name), now renamed *Australopithecus africanus.* This was the start of many remarkable hominid fossil finds at Sterkfontein and environs, and the search continues to this day. In 1997 the world's most complete hominid skeleton, embedded in cave breccia and popularly known as 'Little Foot', was discovered by Dr Ron Clarke in Sterkfontein's Silberberg Grotto. It has now been excavated, bone by bone, and, with a recently determined age of 3.67 million years, represents Sterkfontein's oldest species of apeman, even older than 'Mrs Ples'. So far, more than 750 hominid fossil specimens have been collected from Sterkfontein, and while hominid fossils have been the prime focus, vast numbers of fossils of other, now-extinct vertebrates have also been found, including other primates, large sabre-tooth cats, giant hyaenas, wolf-like dogs, horses and numerous antelope species.

A bronze bust at the exit of the caves of Dr Robert Broom (1866–1951), holding the 'Mrs Ples' skull that he discovered in 1947. He is recognised as the 'father of palaeo-archaeological exploration' in South Africa.

system. There have been several periods of karsting into this ancient dolomitic limestone, perhaps starting some 20 million years ago, although the most important cave development probably started around 5 million years ago.

Rock formations

There are three very different types and ages of rocks in the cave:

- Dolomite: Dark to light grey, bedded dolostone consisting of calcium-magnesium carbonate ($CaMg(CO_3)_2$). In places the dolostone is stromatolitic: these primary accretionary structures were formed in warm shallow water as ultrafine carbonate sediment, chemically deposited by blue-green algae (cyanobacteria), which created vast marine algal beds, akin to today's tropical

Outside of the caves are good exposures of light grey algal laminated and stromatolitic Malmani dolomite (or dolostone), interbedded with more resistant layers of brownish chert.

For palaeontologists who work in the Sterkfontein Caves, it is the young cave breccias that are of the greatest interest, being the only rocks that contain fossils. Above is a section of such breccia, comprising dolomite and chert fragments set in a matrix of what was sand and mud; and it may include bones and plant remains, now all cemented by secondary limestone into a concrete-like mass.

coral reefs. These are now seen as crinkled algal layers and small rounded stromatolites, and also contain numerous thin chert layers (see Sudwala Caves, pp.181–184).

- Cave formations: After slow dissolution by ground water had created solution (karst) cavities in the dolomite, the water table became lowered and air-filled voids resulted. In the dark, humid, empty spaces, lime-rich ground water dripping very, very slowly over millions of years gave rise to a variety of cave formations (stalagmites, stalactites and flowstones) made of pure calcite (see Cango Caves, pp.81–85).
- Cave breccia: Also known as 'cave infill', this is the youngest and only fossil-bearing formation. It started to accumulate in the underground cavities only around 3 million years ago when the caves became open to the surface. Debris consisting of a variable mixture of rock fragments, sand and mud, and including animal remains and wood, was washed in and piled up on cave floors as irregular layers or heaps of sedimentary material – the source of the countless fossils that have been recovered. Over time these crudely layered deposits became cemented by percolating lime-rich water to form concrete-like beds.
- It is in these rocks that fossil bones have been ideally preserved, and from which they are painstakingly recovered. The cave breccias, which have here been grouped into six distinct geological 'members', were deposited over a long time period and age estimates (always debatable) range from more than 3 million years to less than 1.5 million years (Pliocene to Pleistocene). Numerous fossilised animal and wood species have been recovered from the waste dumps of the lime-miners, as well as hundreds of stone tools (also see Makapans Valley, pp.234–238).

The rocky dolomitic soil supports the widespread and very hardy, fibrous-stemmed plant called 'bobbejaanstert' or Xerophyta retinervis, *here blooming in a layer of chert.*

Things to see and do

- **Go on a guided underground trip:** On the tourist route you descend about 40m into a vast underground chamber; here you will marvel at the spectacular cave formations and see the underground lake and fossil-bearing cave breccia. On exiting, you can follow the boardwalk that takes you past the old surface excavations.
- **See the Exhibition Centre:** Housed in the Phillip Tobias Building, this is not to be missed. There are excellent displays showing Earth's geological history, how dolomite caves are formed, and the evolution of life on Earth. You can learn about our very distant apeman relatives who once lived in this area.
- **Visit Maropeng:** Located some 9km northwest of Sterkfontein off the Hekpoort road, this is the official visitors and education centre for the Cradle of Humankind, along with conference facilities. Here the full story is told of how our apeman ancestors are thought to have developed into humans. Exhibitions are held here periodically.
- **Visit Wonder Cave:** This enormous underground cavern is found on higher ground about 6km northeast of Sterkfontein within the Rhino and Lion Reserve. It still contains abundant unmined stalactites and has relatively young deposits of cave infill without hominid fossils. It is open to the public and access is by lift.

Steps at the entrance into Sterkfontein Caves. Visitors are accompanied by an official and accredited guide.

Dr Phillip Tobias, who died in 2012, and whose influence, in recent times, on human origins research, was immense.

As you exit from the underground tour you can listen to the guide telling the story of 'Mrs Ples' and see the bronze busts of Drs Robert Broom and Phillip Tobias, and other commemorative plaques.

44 Walter Sisulu National Botanical Garden

A FAVOURITE OUTDOOR VENUE

A garden that takes you on a walk through geological time

Witpoortjie Falls, at the headwaters of the Crocodile River, is the iconic centrepiece of the botanical garden. Here the stream cascades over ledges of hard, south-dipping Orange Grove quartzite at the base of the Witwatersrand Supergroup.

Situated at the headwaters of the Crocodile River near Krugersdorp, and with iconic Witpoortjie Falls as its centrepiece, this treasured conservation area is a microcosm of the natural environment that once existed along the pristine ridges and valleys of the Witwatersrand. The 275ha area is an ideal outdoor classroom for environmental and geological education, consisting of both landscaped indigenous gardens and wild unspoiled rocky Highveld grassland.

Location and getting there

- The garden is located at the end of Malcolm Road, Poortview Agricultural Holdings, on the West Rand. The best route from Johannesburg is via Hendrik Potgieter Road (M47/R47); from the Krugersdorp-Pretoria highway (R28/N14) turn off at the Pinehaven intersection.
- GPS (entrance to car park): 26°05′14″S; 27°50′42″E

Setting the scene

Witpoortjie Falls is shown on Fred Struben's gold-prospecting sketch map of 1885 but, discouraged by rather weak gold findings in the area, Fred and his brother Harry moved eastwards. These Witwatersrand hills remained largely undeveloped and, in 1982, the property was acquired for development as a national garden, which first opened in 1986 as the Witwatersrand National Botanical Garden. The name 'Witwatersrand' was given to the area by early Boer settlers, probably because of the pale (whitish) quartzite ridges that were the source of numerous streams, hence the 'white ridge of waters'.

The unique Geological Display Garden provides a walk through South Africa's geological time, showing a variety of important regional rock types, with easy-to-understand explanations, and geological ages are marked on the walkway.

A large specimen of hard, weathered Orange Grove quartzite, intruded by a whitish quartz vein, and showing a polished 'window' and explanation.

Regional geology

The formations that make up the rocky hills are part of the **West Rand Group**, which is not host to any significant gold deposits, and forms the lower division of the 7,000m-thick **Witwatersrand Supergroup**. Underneath are much older greenstones of the Archaean-age **Roodekrans Complex**, which, together with granite, formed the floor of the shallow Witwatersrand Sea onto which sediments were deposited from about 2,950 million years ago.

Different varieties of granitic rock that are found in the Gauteng region. All the explanatory plaques are made of polished Bushveld Red granite.

Landscape and landforms

The high ground that forms the Witwatersrand Plateau is a remnant of the very old land surface predating the break-up of Gondwana.

Above the waterfall you can see south-dipping, faulted Orange Grove Formation quartzite and part of the Parktown Formation shales, these being the lowermost formations of the Witwatersrand Supergroup. The white dashed lines represent the top and bottom of the Orange Grove quartzite and the black line shows the almost vertical fault (f) that cuts the strata (relative movement indicated).

Right: A bedded Malmani dolomite specimen interlayered with chert. Inset: The small polished 'window' clearly reveals the crenulated algal laminations and a small stromatolitic structure.

A polished slab of Ventersdorp basaltic lava showing countless amygdales – originally bubbles in the very liquid lava that later became infilled by whitish minerals.

Weathered serpentinite that comes from the Archaean Roodekrans ultramafic complex exposed at the base of the waterfall, and now altered to talc-rich greenstone.

To the north of the Witwatersrand ridge, at a lower elevation, is an undulating plain of older granitic rocks of the Johannesburg Dome. South-dipping sedimentary formations have created the characteristic 'ridge-and-valley' topography of the Witwatersrand, and it is also an important regional watershed, separating streams that flow south into the Vaal River from those that flow north into the Limpopo River.

Rock formations

- Orange Grove quartzite: This forms the prominent northwest-trending ridge that is typical of the Witwatersrand's northern edge. This was originally clean sand eroded from granitic mountains and deposited into the early Witwatersrand Sea.
- Over immense time and through deep burial, the sand became compacted and metamorphosed to form very hard rock.
- Iron-rich Parktown shale: This lies above the quartzite near the top of the ridge, and also includes thin beds of quartzite. It weathers to a reddish-brown but is dark and slate-like when fresh, and was deposited in deeper, quieter water.
- Brixton quartzite: Southwards, in the distance, is the next quartzite formation of the Witwatersrand, on top of the Parktown shales.
- Greenstone: These rocks are difficult to see because they outcrop poorly but they can be seen as soft, slippery grey-coloured talc schist outcrops beneath the quartzite close to the start of the steep trail to the top of the waterfall. They represent ultramafic rocks that are some of the oldest rocks in South Africa.

The Strubens and the Confidence Reef

Fred Struben, backed by brother Harry, a successful businessman from Pretoria, started systematic gold prospecting along the hills of the West Rand in 1883 before Johannesburg existed. He found and recorded numerous traces of gold on his 1885 map of the area, and some were undoubtedly from pebble conglomerates. But it was Fred Struben's quartz reef discovery on the farm Wilgespruit in late 1884 that persuaded them they had struck it rich. It was named the Confidence Reef and samples were shown by Harry Struben to the Volksraad in 1885. A small mine developed and a stamp mill was imported from England, but gold recovered was disappointing. This historic gold mine lies within the Kloofendal Nature Reserve, some 7km to the southeast of the Garden and is a Provincial Heritage Site. The Strubens' mine and mill attracted many prospectors and entrepreneurs to the area, and almost certainly gold-bearing Witwatersrand conglomerate was crushed here even before the discovery of the Main Reef at Langlaagte in February 1886. Considerable gold prospecting was done by the Struben brothers and some believe that they are the true discoverers of the Witwatersrand goldfield.

Things to see and do

- **Walk through the Geological Display Garden:** This beautifully landscaped section was formally opened in 2008. Using extremely large specimens it provides a magnificent open-air display of 15 of the country's most important geological formations, well described in easy-to-follow language on polished rock plaques. The displays are cleverly laid out along a meandering geological time-line from oldest to youngest, starting at 3,500-million-year-old greenstone and ending at 3.5-million-year-old, bone-bearing cave breccia, and they tell much of the story of South Africa's interior geology.
- **Hike the JCI Geological Trail:** It starts at the base of Witpoortjie Falls and follows a roughly circular, 2.4km route up the cliff next to the waterfall, westward along the ridge from where panoramic views can be enjoyed, and returns via a scree-covered slope rich in proteas. You will see:
  - Poorly exposed outcrops of slippery talc schist at the start of the trail near the waterfall.
  - Cliffs of massive Orange Grove quartzite, the rocks that form the prominent northern scarp-face.
  - A descriptive plaque at the top of the waterfall that illustrates the geology across the valley.
  - Outcrops of reddish-brown Parktown shale, including the strongly magnetic Water Tower shale.

The paved pathway that leads to the iconic Witpoortjie Falls and the Witwatersrand ridge.

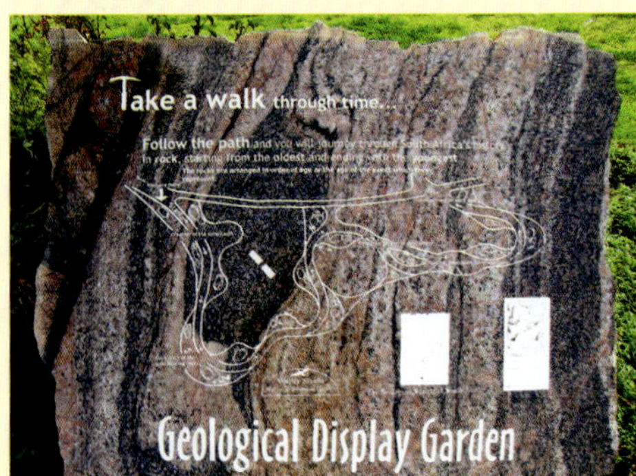

At the start of the Geological Display Garden a rock plaque of Vredefort granitic gneiss (a once popular decorative stone) shows a plan of the area. The inscription reads: 'Follow the path and you will journey through South Africa's history in rock, starting with the oldest and ending with the youngest'.

45 Gold Reef City No.14 Shaft

A RELIC OF JOHANNESBURG'S GOLDEN PAST

The only authentic Witwatersrand underground gold mine that can be visited by the public

At the entrance to Gold Reef City is a historic once steam-powered, 10-stamp rock crusher. In the early days of mining on the Witwatersrand this was the way the gold-bearing ore was crushed before the resulting slurry was passed over mercury amalgam plates to recover the gold. By 1915 over 10,000 such stamps were installed on the Witwatersrand.

Gold Reef City theme park was started in the 1980s as a 'working replica' of the Witwatersrand's gold-mining industry, and is centred round No. 14 Shaft at Crown Mines, at one time the world's richest and largest gold mine. Underground mining has long since stopped but this historic shaft continues to operate as a tourist attraction – a walk down memory lane – and part of a mining heritage tour.

Location and getting there

- The theme park is located in southern Johannesburg, in an urban landscape that was once characterised by mine headgears, tailings dumps and blue-gum plantations. The park lies about 5km southwest of the city centre and adjacent to the M1 highway. To get there take the Booysens Road off-ramp (M27) and follow the signs along Northern Parkway Road.

- GPS (entrance gate): 26°14′13″S; 28°00′43″E

Setting the scene

The discovery in early 1886 of the gold-bearing Main Reef conglomerates on the farm Langlaagte, about 4km northwest of Gold Reef City, was the most momentous mining event in South Africa's 19th century history. Gold was soon being mined all along the shallow, weathered conglomerate beds or 'reefs', as the gold-bearing rock was called. Numerous small gold mines were working between Langlaagte and the new mining town of Johannesburg, and by 1909 the amalgamation of 12 of these operations resulted in the creation of a new company, Crown Mines Limited. Deep drilling proved that the south-dipping reefs continue far to the south, and underground mining would require deeper shafts.

No. 14 Shaft at Crown Mines was sunk to an initial depth of 980m in 1916 and was equipped with this steel headgear and winder. It is still in operation today, taking tourists down to an upper level of the old mine.

Regional geology

Johannesburg is underlain by the legendary **Witwatersrand Supergroup**, an approximately 7,000m-thick sequence of predominantly sedimentary rocks ranging from around 2,950–2,700 million years in age. The sequence

The story of Crown Mines

Crown Mines Limited was the world's first super-mine and at one time the busiest and most profitable gold mine in the world. Peak production was in the 1930s and, when it closed down in 1977, this mine had produced over 45 million troy ounces (over 1,400,000kg) of gold from production over nearly 90 years. As mining followed the ever-deepening conglomerate reefs, new shafts had to be sunk to access the deeper reefs, and one of these was No. 14 Shaft, the third generation of surface shafts, completed in 1916 to a final depth of 980m. Mine workings continued far below No. 14 Shaft, accessed by later sub-vertical (a vertical shaft that starts below surface) and sub-incline (an angled shaft that starts below surface) shafts, as the south-dipping Main Reef conglomerates were followed to 2,875m below the surface, which is well below sea level. Today there is little evidence on the surface to tell of the hundreds of kilometres of underground tunnels or of the many shaft headgears that once dotted the landscape.

Figure 33: A simplified north-south geological cross-section through Crown Mines Eastern Shaft System, showing the location of No. 14 Shaft. At Gold Reef City visitors can go underground to the shallow Kimberley Reefs area. It is notable that the old workings went to well below sea level.

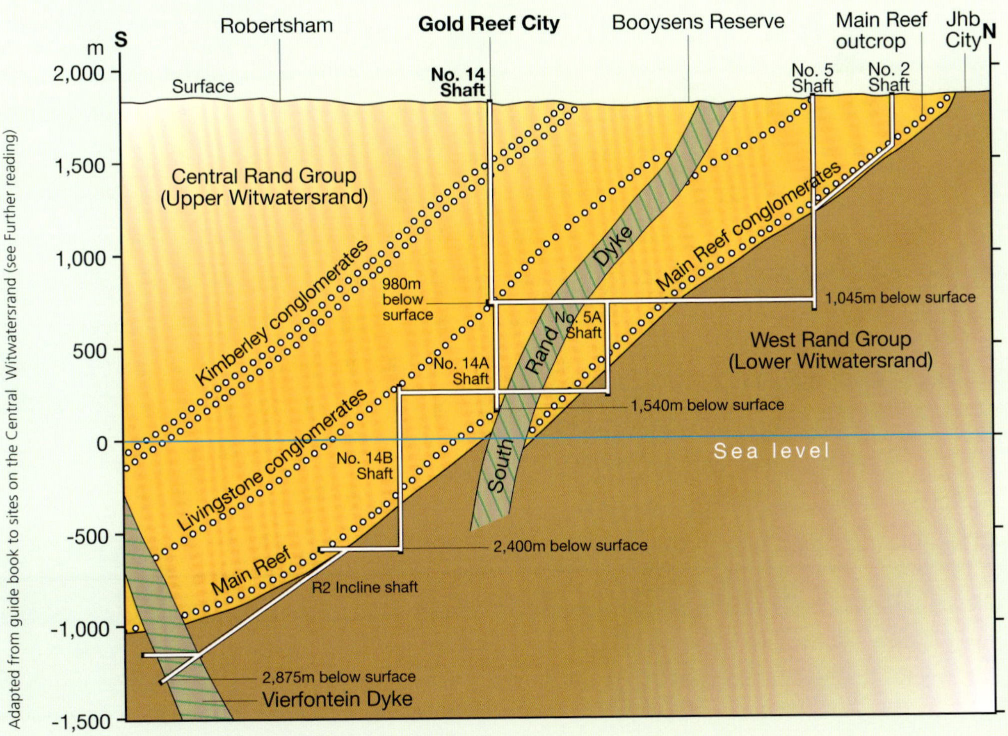

is made up of the lower **West Rand Group** (see Walter Sisulu Botanical Garden, pp.276–280) and the upper, gold reef-bearing **Central Rand Group**, where Crown Mines is situated. The upper group formations were deposited more or less horizontally as clean sand and pebble beds around the margin of the shallow Witwatersrand Sea, but they now dip around 35° to the south, at least partly the result of the Vredefort meteorite impact (see pp.154–161).

A recreated 'period' building, now used as a hotel at Gold Reef City.

Recent open-cast mining operations immediately south of George Harrison Park, the discovery site of the Main Reef conglomerates by prospector George Harrison in 1886.

The iconic (labelled) exposure of the south-dipping thick Main Reef and thin Main Reef Leader at George Harrison Park, a Provincial Heritage Site at the old Langlaagte gold mine. Inset: A close-up view of outcropping pyritic Main Reef quartz pebble conglomerate as exposed at George Harrison Park.

Landscape and landforms

Geographically, the Witwatersrand forms an elevated plateau that reaches almost 2,000m above sea level and is a regional watershed separating the Vaal and Limpopo river systems. Before the discovery of gold in 1886 it was sparsely populated, and comprised rocky hills, flat grassy plains and waterfalls. It was on this flat, treeless Highveld landscape that the first gold mines of the Witwatersrand were established, and where Johannesburg came to life. The mines have now all been closed and most mining structures have been dismantled. Many of the iconic man-made tailings dumps have been reprocessed to recover small amounts of residual gold and the sand tailings sent away to be dispersed further afield.

Rock formations

- At Gold Reef City there is no geology to be seen on the surface but a trip underground takes you into south-dipping Kimberley quartzites (see the geological cross-section through No. 14 Shaft, p.283).
- In the upper levels at Crown Mines some of the small-pebble conglomerate beds, known as the Kimberley Reefs, were mined.
- A large unmined gold resource still exists – its gold content having been considered too poor in the old days, with the deeper Main Reef providing richer, more continuous reef.
- The sought-after Main Reef conglomerates outcropped nearly 3km to the north, and are an eastward extension of the discovery outcrop at Langlaagte (pictured above).

The Witwatersrand gold rush

In the early 1880s gold prospecting had gripped the old Transvaal Republic, and wonderful discoveries were being made around Barberton (see Barberton-Makhonjwa Geotrail, pp.166–170). The story goes that one Sunday morning in February 1886, while working on a farm called Langlaagte, an itinerant prospector and builder by the name of George Harrison (and possibly also his colleague George Walker) crushed some good-looking weathered pebble conglomerate – and found a rich tail of gold in the pan. The outcrop of the fabulous Main Reef had been discovered. Soon this find was common knowledge and in July 1886 Harrison officially reported to the authorities his discovery of 'a payable goldfield'. The biggest gold rush of all time was on, and Johannesburg's future was assured (see George Harrison Park below).

Things to see and do

- **Do 'Jozi's Story of Gold' Heritage Tour:** This guided historical tour around Gold Reef City is a must-do activity and the trip down No. 14 Shaft provides an authentic underground mining experience. The tour includes the history of the discovery of gold and how Johannesburg came about.
- **Visit George Harrison Park:** This National Monument, now a Provincial Heritage Site, is located about 5km west of the city centre off Main Reef Road (R41). It was around here that South Africa's gold-mining history was indelibly made when the gold-rich Main Reef outcrop was found (see text box above).
- **See the Ferreira Mine stope:** In 1986, during excavations for Standard Bank's new head office at 5 Simmonds Street in central Johannesburg, an old hand-dug tunnel of the Ferreira Gold Mine dating from the 1890s was accidently discovered. The mine was founded by Colonel IP Ferreira, one of the pioneering diggers and a mining camp organiser, and this historic underground digging has been preserved as a fascinating small museum.

Exposure of the south-dipping Main Reef conglomerates at historic George Harrison Park. In the background is the recent open-cast mining further down the dip of this outcrop.

Standard Bank

A preserved stope of the old Ferreira Mine located deep under the Standard Bank's Head Office at 5 Simmonds Street in central Johannesburg.

NORTHERN CAPE

OVERVIEW OF GEOLOGY AND LANDSCAPES

This final chapter focuses on Geosites 46–50, which are shown on the national maps (pp.12–13); there are no regional maps in this section as the region is vast compared with the sprinkling of geosites. Covering much of northwestern South Africa, and largely dry to semi-desert, this region reveals well-exposed and, in most places, complex geology, mostly dating from the Proterozoic. Although geological sites are not featured in all the major formations, certain formations are included here for geological interest. The landscape is divided into several major geomorphic domains and, in terms of the geosites discussed, the following are important: the Ghaap Plateau in the east; the valley of the Lower Orange River; the Namaqua Highlands in the west; and in the south, the plateau of the Upper Karoo.

The major geological formations of the region are listed below:

■ Transvaal Supergroup
(Late Archaean to Early Proterozoic)

This major sequence is also present in the eastern part of the Northern Cape within the Griqualand West Basin, which has the same age and character as the Transvaal Basin. From around 2,650 million years ago it was part of the same shallow continental Transvaal Sea that covered much of the Kaapvaal Craton. Here the **Vryburg Formation** is the equivalent of the Black Reef quartzite, and is followed by the **Ghaap Group**, comparable to the **Chuniespoort Group**; it includes dolostone formations of the Campbell Rand dolomite, the Northern Cape's twin of the Malmani dolomite. Stratigraphically above are sedimentary and volcanic formations of the **Postmasburg Group**, which is correlated with the lower part of the **Pretoria Group**.

● **Campbell Rand Subgroup** (lower part of the **Ghaap Group**)**:** This sequence of carbonate rocks is exposed on the extremely flat Ghaap Plateau that lies between Vryburg and Kuruman, and extends southwards to the Orange River. Estimated to be at least 1,600m thick, the sequence consists predominantly of horizontally layered and stromatolitic dolostone, essentially the same as the algal formations of the Malmani dolomite. Its origin is explained on p.163, including its unique evolutionary photosynthesising, oxygen-producing capability. The plateau is not crossed by any rivers and is characterised by rainwater infiltration and subsurface drainage. Numerous linear dykes have separated the flow of underground water into large compartments, leading to the formation of springs, the best-known being in the west at the Eye of Kuruman (**Geosite 47**).

● **Postmasburg Group:** Going westwards, this overlying group forms an important divide in the topography, and makes the ridges of the Southern Kalahari geomorphic province. Known as the Kuruman Hills and further south as the Asbestos Mountains, they hold vast resources of banded iron formation and manganese, as well the infamous blue asbestos (crocidolite) that was once exploited here.

■ Olifantshoek Supergroup
(Middle Proterozoic)

The prominent ranges of the Langberg in the south and Korannaberg in the north are part of this thick sequence, and form part of the arid landscape of resistant ridges and flat sandy valleys typical of the Southern Kalahari geomorphic province. The formations are predominantly made of quartzites, originally deposited as sandstones along the western edge of the Kaapvaal Craton about

1,900 million years ago. Later, these formations were deformed by folding and thrusting that came from the west during the period of Kheis mountain building. They are correlated in time with the red beds of the Waterberg and Soutpansberg Groups (see p.243).

Namaqua Metamorphic Province
(Middle Proterozoic)

This immense and extremely complex geological province extends over about 100,000km^2, from east of Upington to the Atlantic coast, and northwards into southern Namibia. It is also part of the strongly metamorphosed sequence that continues eastwards under Karoo cover and reappears in KwaZulu-Natal as the **Natal Metamorphic Province** (see p.114). Formations consist of a wide variety of granites and gneisses, and of sedimentary rocks that were tectonised and intruded 1,200–1,000 million years ago along the southern and eastern margin of the Kaapvaal Craton during the major Namaqua mountain-building event. The landscapes that have developed are vast and, in places, spectacular. Of interest here is the valley of the Lower Orange River, which hosts the Augrabies Falls (**Geosite 48**); and the Namaqua Highlands near Springbok, close to the western part of the Great Escarpment, where South Africa's first copper prospect exits as the old Koperberg Mine (**Geosite 49**).

Karoo Supergroup
(Carboniferous to Jurassic)

Part of the western Northern Cape is underlain by unfolded **Dwyka** and **Ecca Groups**, within the Main Karoo Basin. The area is well known for its many large pans such as Grootvloer and Verneukpan near Brandvlei. In the south, alluvial flood-plain sediments of the lower **Beaufort Group** are exposed and are world renowned for abundant fossil evidence of prehistoric reptile life, as revealed at the intriguing Fraserburg Fossil Surface (**Geosite 50**). The landscape grades from the Bushmanland Plain in the north to the Upper Karoo in the south, followed by the rugged Nuweveld Mountains where intrusions of **Karoo dolerite** have played a dominant role in the formation of the southern Great Escarpment (see p.309).

Kimberlite intrusions
(mostly Jurassic to Cretaceous)

There are reported to be around 850 known kimberlite intrusions in southern Africa, of which only about 50 are diamond bearing, and very few contain diamonds in economic quantities. Nevertheless it was around what is now Kimberley that, from 1871 onwards, several pipe-like volcanic intrusions were mined for their diamonds, sparking the biggest diamond rush of all time. Once known as Colesberg Kopje, this first known pipe was mined to become Kimberley's Big Hole (**Geosite 46**), and was the richest of a cluster of several kimberlites. Over the course of many millions of years, erosion removed the upper parts of the volcanic pipes and countless almost indestructible diamonds were shed into streams and rivers, where they are still being mined as alluvial diamonds; or they found their way to the Atlantic coastline, where they are recovered in abundance along the old beach deposits and out at sea.

The historic steel girder and stone bridge dating from 1884 that crosses the Vaal River near Barkly West, the site of South Africa's first major river diamond diggings at Canteen Koppie and the notorious Klip Drift diamond rush in 1870.

46 Kimberley's Big Hole

WORLD'S LARGEST HAND-DUG HOLE

This now mined-out kimberlite pipe produced over 14.5 million carats of diamond

De Beers

The world-famous Big Hole, as seen from the viewing platform on the western side. The geology in this upper part of the excavation is clear: the thin brown top layer is surface debris atop a weathered dolerite sill, followed by a thick layer of blueish Karoo (Ecca Group shale) and thin Dwyka Group tillite that overlies the much harder solid vertical wall of Ventersdorp Supergroup lava (see cross-section on p.291). The water level has settled at around 175m below the land surface.

In July 1871 a member of Fleetwood Rawstone's 'Red Cap Party' found an 85-carat diamond on the small hill that was known by diamond diggers as Colesberg Kopje. This chance discovery sparked the biggest diamond rush the world has ever seen, an event that was to change South Africa's fortunes forever. The hill was excavated and became the famous Kimberley Mine, one of five large diamond mines in the area operated by De Beers Consolidated, and once the richest diamond field in the world. Originally known as New Rush, the town that rapidly established was renamed Kimberley in 1873 after Lord Kimberley, then Secretary of State for the Colonies.

Location and getting there

- The Big Hole is in the centre of Kimberley, which grew up in response to major diamond-mining activity in the area. To visit the Big Hole, you need to get onto West Circular Road, from where the Visitors Centre complex is well signposted.

- GPS (viewing platform): 28°44′21″S; 24°45′22″E

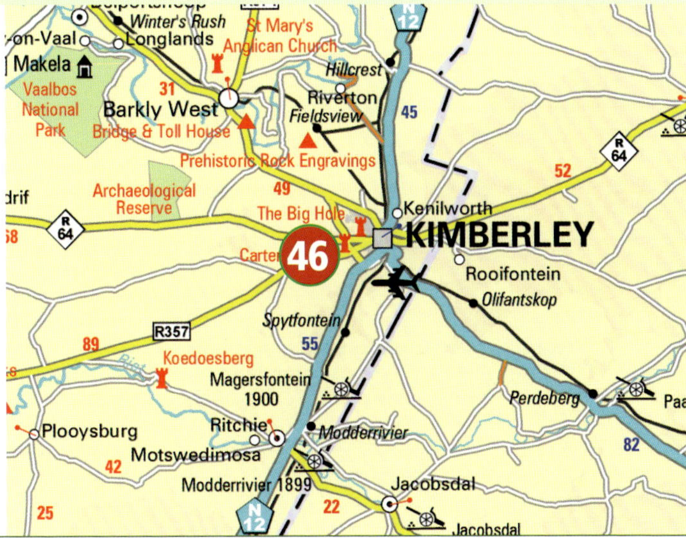

Setting the scene

The discovery of diamonds at Colesberg Kopje on the De Beer brothers' farm (hence the company name) led to a mad scramble for claims and within days 800 claims had been staked; soon more than 2,000 hopeful prospectors were feverishly at work. The diggers frenetically attacked the 'yellow ground' of the hill and it rapidly disappeared to become the opencast working of the Kimberley Mine, now the Big Hole. 'Blue ground' (fresh kimberlite) was reached at a depth of around 20–30m and it was soon realised that the diamonds came from the kimberlite pipe that continued downwards.

De Beers

Fleetwood Rawstone's 'Red Cap Party' seen sieving gravel in 1871. Their cook, Damon Esau, is said to have found the first diamond on Colesberg Kopje in July that year. What was once a low hill soon became the awesome Big Hole.

De Beers

Colesberg Kopje soon disappeared as mining progressed ever deeper. By 1873 diggers reached their claims using a confusing system of winders, cables and buckets, and mining became difficult and dangerous. Notice that some claims blocks have not been fully worked.

A suspended walkway that extends from the Visitors Centre out over the western side of the excavation, and provides a superb view of the Big Hole.

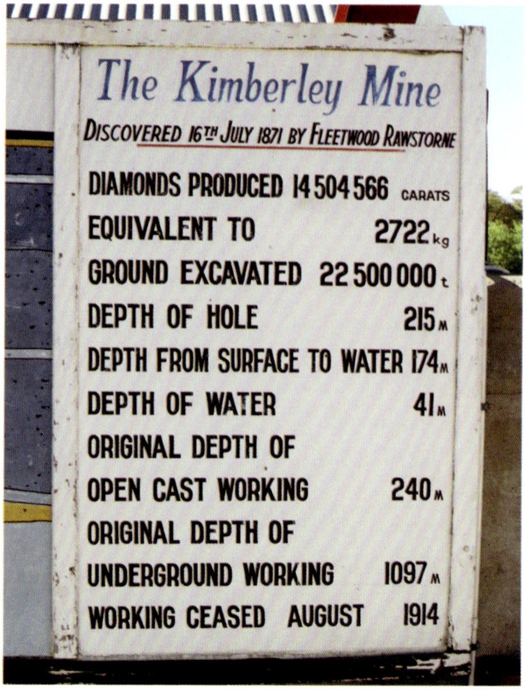

This information board, located at the start of the suspended walkway, provides visitors with the essential statistics of the old Kimberley Mine that produced 14.5 million carats of diamonds up to its closure in 1914.

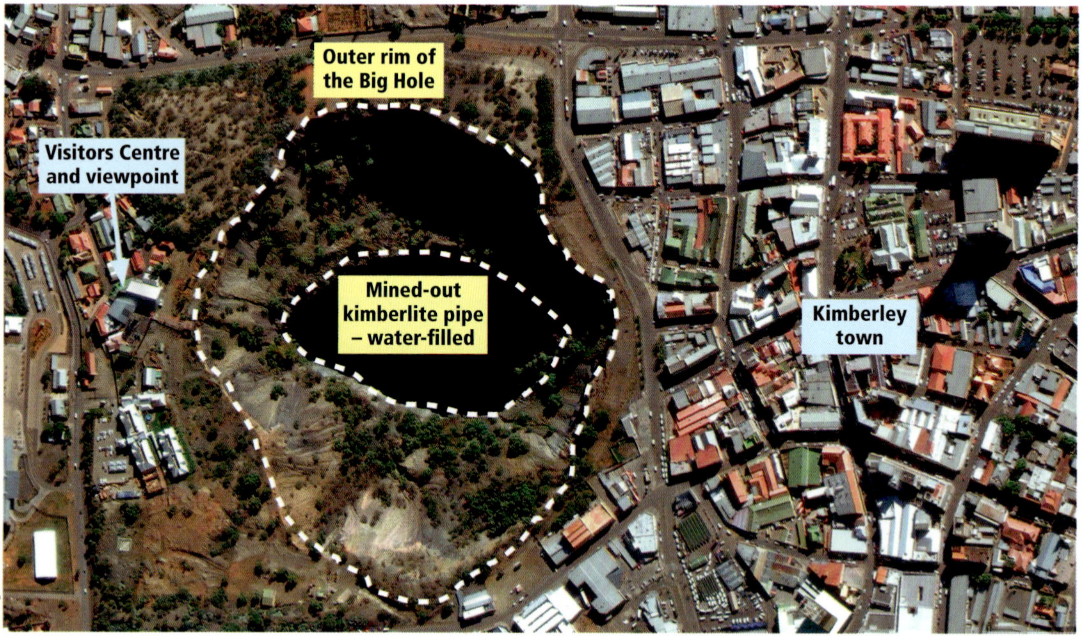

Google Earth Pro

A Google Earth image of the Big Hole located in the centre of Kimberley. Notice the outline of the kimberlite pipe in the centre (and partly in shadow) where it punched through hard Ventersdorp lava. The pipe is actually elongated east-west, and is much smaller than the roughly circular excavated area at surface. This is the world's largest hand-dug hole.

Regional geology

The **Kimberley Pipe** is one of several diamond-rich volcanic pipes and smaller intrusions that cluster around Kimberley, part of a volcanic event that took place during the mid-Cretaceous. The kimberlite intruded explosively through the **Ventersdorp** and **Karoo Supergroups**. Of critical importance is that the Kimberley region is underlain by thick continental crust of the ancient **Kaapvaal Craton**, which is key to the genesis of diamonds. Diamonds form at a depth of more than 150km in the upper mantle, but only within the relatively cooler diamond-stable zone that exists beneath thick cratons. The diamonds are brought to the surface as accidental inclusions in kimberlite magma of even deeper mantle origin.

Figure 34: An east-looking cross-section through the Kimberley Pipe, now seen as the 'Big Hole'.

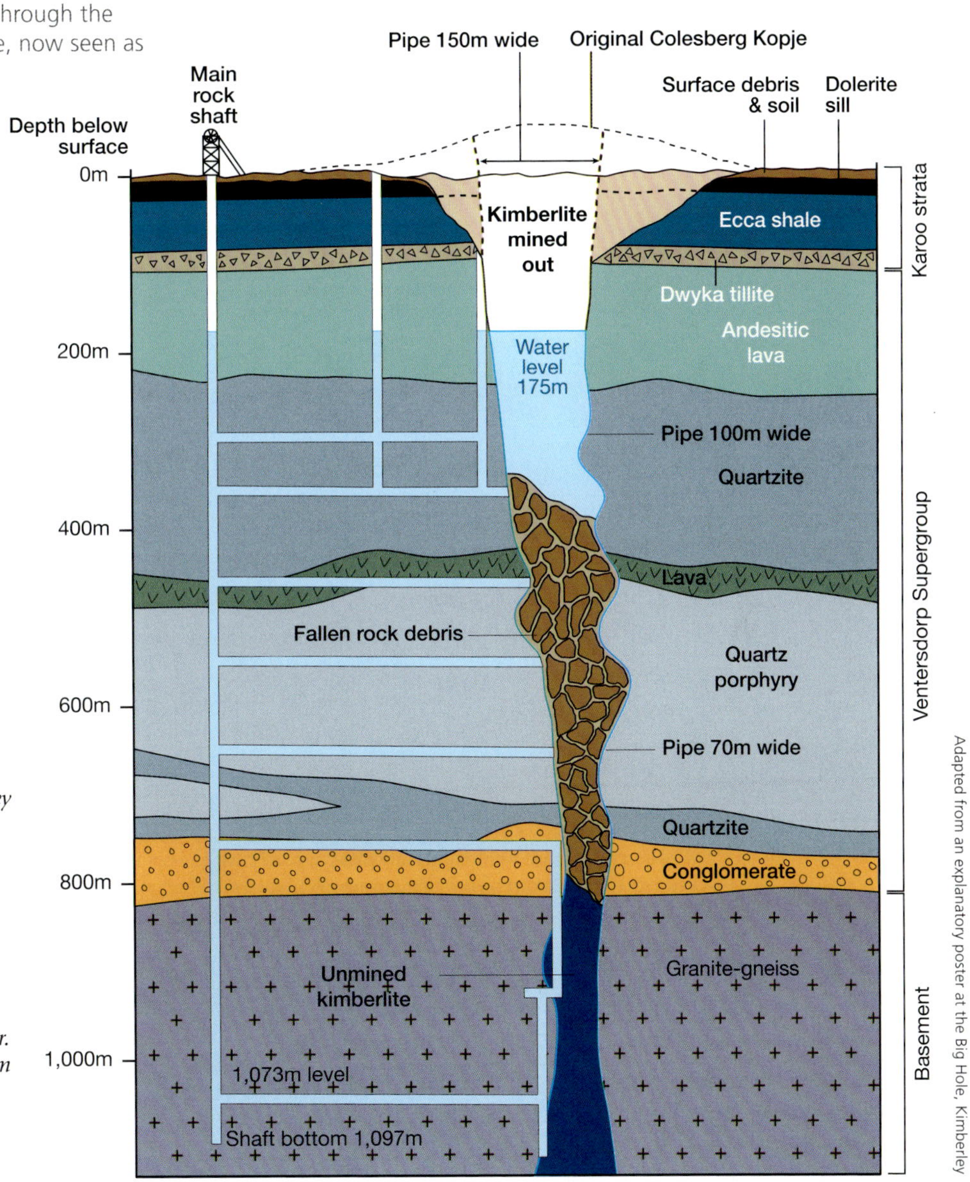

A schematic north-south geological cross-section through the old Kimberley Mine and the surrounding rock formations. Mining from surface stopped at 240m below surface, when underground mining took over. The shaft bottom was 1,097m below surface when the mine closed in 1914.

The Kimberley Pipe

This is one of the world's most famous kimberlite pipes, the smallest of five large diamond-bearing pipes clustered around Kimberley. Roughly oval in shape, and typically carrot-shaped, it narrows as it goes down, from about 300m by 150m at the surface to around 170m by 70m at a depth of about 650m (**Figure 34**). This is the lower 'diatreme' portion of the pipe, and as it descends the narrowing pipe connects with an east-west-trending kimberlite dyke, the probable feeder fissure. This kimberlite is about 87 million years old; the diamonds that it contained are very much older.

In his classic 1914 publication, *The Diamond Fields of Southern Africa*, geologist Percy Wagner wrote: 'The mighty vent now stands open to a depth of 1,100 feet and ... shows clearly that it has been formed by coalescence of at least three distinct pipes. These are arranged along a fissure having an east-northeast-west-southwest trend. The fissure is well exposed on the eastern wall of the open mine where it is seen to be occupied by a narrow dyke of kimberlite ...'

The pipe was dug by hand to a depth of 240m using picks, shovels and explosives to create the Big Hole. Thereafter it was mined underground by a shaft system that reached 1,097m below the surface by the time the mine shut down in 1914. Some 22.5 million tons of rock were removed, from which over 14.5 million carats (2,722kg) of diamonds were recovered. Both the grade (the quantity) and the quality (clarity and colour) of diamonds from this pipe are said to have been higher than those from the other Kimberley mines.

Landscape and landforms

At surface the Big Hole is very roughly circular and about 500m across. Considerable slumping of softer sedimentary side walls into the hole has taken place, but the elliptical outline of the kimberlite pipe is clear in the deeper and harder Ventersdorp lava. In the Cretaceous the explosive eruption of the Kimberley volcano probably formed a massive crater into which sediment was deposited and in which a variety of life existed (see artist's impression, right). The upper part of the pipe was removed during many millions of years of ongoing denudation of South Africa's interior plateau, which here amounted to the removal of an estimated 900m of overlying Karoo formations.

Maggie Newman and De Beers

Artist's impression of an eroding kimberlite crater and lake during the Cretaceous, long after eruption. The distant mountains show retreating Drakensberg lavas and Clarens sandstone overlying Karoo sedimentary rocks. At that time dinosaurs were dominant and coexisted with small mammals, insects and early flowering plants.

Rock formations

- Kimberlite is the host rock to almost all natural diamonds. It is an unusual and heterogeneous ultramafic rock formed in a volcanic vent (or pipe), coming from great depths beneath the Earth's crust and bringing up samples of upper mantle rocks.
- When weathered and soft near the surface, the rock was known by diggers as 'yellow ground' and when fresh and harder, deeper down, it was called 'blue ground'.
- The Kimberley Pipe, a volcanic diatreme, is now called 'Group 1 kimberlite', a dark, blue-grey intrusive rock, packed with wall-rock inclusions.
- Kimberlite rock was named after the town of Kimberley by Professor Henry Carvill Lewis in 1886, and is now found worldwide.

Things to see and do

- **Visit the Big Hole:** At the Visitors Complex you can learn about diamonds, mining and the history of Kimberley, providing a variety of experiences and a look into the 175m-deep Big Hole. Outside of the museum unique kimberlite 'mantle nodules' or xenoliths can be seen.
- **Go to Barkly West (the Canteen Koppie diggings are a Provincial Heritage Site):** Some 35km north of Kimberley along the Vaal River, major alluvial diamond mining took place here at Klip Drift in 1870. The small museum illustrates the historical diggings.
- **See the Nooitgedacht glacial floor (Provincial Heritage Site):** Some 20km north of Kimberley and 6km off the R31 (easily accessed) there is a glacial pavement on worn-down Ventersdorp lava, showing movements of Dwyka ice-sheets and it also displays numerous San petroglyphs (rock engravings).

De Beers

This is what it's all about – the very rare occurrence of a crystalline diamond in a piece of kimberlite.

Some of the recreated buildings of old Kimberley and a working tram-car at the Visitors Centre.

47

Eye of Kuruman

PROVINCIAL HERITAGE SITE

The strongest natural spring in the southern hemisphere and a welcome oasis in the dry Kalahari

The prolific Eye of Kuruman, located within in a small dolomite outcrop, runs into a large, lily-covered pond that is home to a variety of large fish. More than 20 million litres of clear water gush out every day at this desert oasis.

More than 20 million litres of pure water gush out of the ground every day from the Eye of Kuruman – a reliable and never-ending supply of water for the community, even in times of drought. It is known as *Die Oog* in Afrikaans and *Gasegonyana* in Setswana.

Location and getting there

- The Eye is located within a pleasant park in the town of Kuruman. Run-off from the spring leads into a large, clear pond. Turn off the N14 main road at Fontein Street for the park's entrance and parking.
- GPS (at The Eye): 27°27′50″S; 23°26′11″E

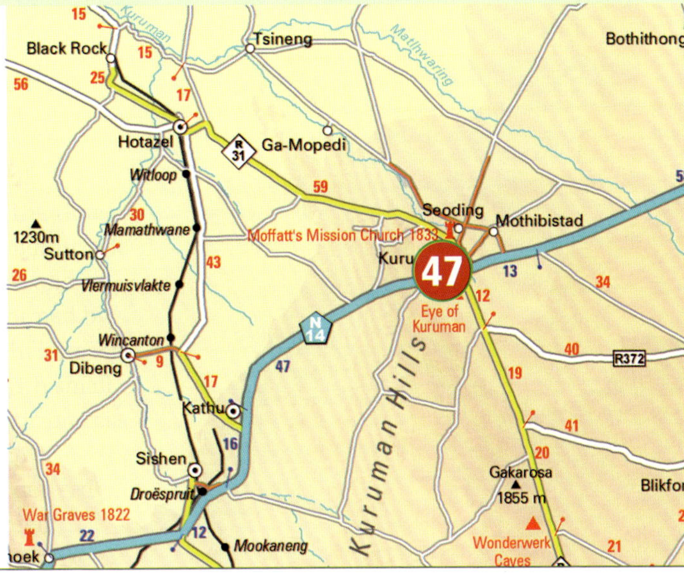

Setting the scene

In 1821 Scottish missionary Robert Moffat and his wife Mary were sent by the London Missionary Society to establish a station in Kuruman, an extremely remote settlement chosen because of its permanent and prolific spring, the only reliable source of fresh water in the entire region. Here the Moffats worked for some 50 years in what was to become the most successful mission station in southern Africa. The spring was not only a key factor in the establishment of the mission station but it also meant Kuruman became an important stop-over for many early adventurers on their journeys of exploration into the interior.

Walter Knirr/Images of Africa

The original church built by Scottish missionary Robert Moffat at the Kuruman Mission Station is still being used today, and is known as 'the fountain of Christianity in southern Africa'.

Ariadne van Zandbergen

A bronze bust commemorates the bearded Moffat who spent some 50 years at the mission.

It is from this inconspicuous, fern-covered cleft in Campbell Rand dolomite that the Eye of Kuruman produces the vast and constant flow of crystal-clear water that is essential for the local community.

Regional geology

The Eye is found in the upper part of the **Campbell Rand Subgroup** of the Griqualand West Basin, the western basin of the **Transvaal Supergroup**, and in rocks that are the equivalent of the Malmani dolomite. These rocks were formed around 2,500 million years ago, as chemical sediments deposited in warm, shallow marine conditions. They formed with the intimate assistance of algal life forms (now seen as stromatolites) and were part of the early Earth's great oxygen generator through the process of photosynthesis (see Sudwala Caves, pp.181–184).

Landscape and landforms

Kuruman is located on the northwestern edge of the Ghaap Plateau, a remarkably flat limestone plain dating from the African erosion cycle, and close to the north-south-trending, ironstone-dominated Kuruman Hills. The surface overlies a zone of ancient karstic solution cavities and channels – the makings of a good aquifer. The Eye lies in the upper Kuruman River drainage, at a regionally low elevation of 1,301m.

Rock formations

- Dark grey, fine-grained dolomite rock (or dolostone) is the only rock seen, dipping to the west at about 10°.
- Although not obvious on the surface, the Ghaap Plateau is crisscrossed by many vertical dolerite dykes, mostly oriented roughly north-south and northwest-southeast. At Kuruman there is also an east-northeast-trending dyke parallel to the N12 highway.
- These dykes result in large hydrogeological compartments being formed in which ground water collects (see text box opposite).
- There are several other springs in the region, including one about 5km to the northeast and known as the Second Eye, which has recently ceased to flow.

How the Eye of Kuruman works

Earlier ground-water measurements revealed that The Eye flowed at a remarkable 18–21 million litres (18,000–21,000m^3) per day. More recent flow figures indicate a higher average flow of nearly 30 million litres per day but most of this is now pumped out by town boreholes 'upstream' from The Eye for municipal use. It is considered to be one of the best springs in the world, and the flow from this natural wonder hardly falters. The essential mechanism is as follows:

- Long after the formation of the ancient dolomite, it was crisscrossed by numerous dolerite dykes. These dykes form physical barriers to the movement of ground water.
- The dolerite dykes divided the thick, extensive dolomite strata into large, somewhat rectangular ground-water 'compartments' that in effect became large ground-water reservoirs, which are periodically re-charged by rainfall (about 445mm average per year at Kuruman).
- Within the dolomite, ground water is present within countless solution cavities and fissures of a widely developed karstic system that formed long after the dolerite dykes were intruded.
- The dolerite compartments slowly leak ground water into adjoining ones where the water table is at a lower level, and from higher elevation compartments to those that are lower.
- Kuruman, alongside the usually dry Kuruman River, is located in a topographic low, so ground water from the Ghaap Plateau to the east and the Kuruman Hills to the west flows towards the town.
- The Eye is located in the northeast corner of a dyke-bounded ground-water compartment. Gravitating ground water is 'pushed' to surface at the lowest point, forming The Eye.

Things to see and do

- **See the Eye of Kuruman:** This is a must-see site, a remarkable place in the arid surroundings. See the constant stream of fresh water issuing from a cleft in the partly overgrown, fern-covered dolomite, flowing into a picturesque pond filled with a variety of fish and other aquatic life. Some believe that the spring is sacred and has healing properties.
- **Visit Robert Moffat's mission station:** No visit to Kuruman is complete without visiting what is often described as the 'fountain of Christianity in southern Africa', about 5km north of the town. This was the starting point for many early expeditions into the interior, including those by David Livingstone. Still a functioning mission station, it is a reminder of the region's impact on local history.
- **See Wonderwerk Cave:** About 43km south of Kuruman on the road to Daniëlskuil off the R31 is the massive Wonderwerk Cave, a National Heritage Site and archaeological dig. The cave extends inwards for 139m and is up to 17m wide. It was formed as a solution cavity in the dolomite strata – much as the underground aquifers around Kuruman were formed.

Ariadne Van Zandbergen

Wonderwerk Cave has recently been recognised as a National Heritage Site because of its immense archaeological value.

48 Augrabies Falls

A SOUTH AFRICAN NATIONAL PARK

Ranked as one of the major waterfalls in the world, its name comes from the original Khoi word meaning 'a place of great noise'

South African Tourism

The Main Falls, seen looking downstream, where the entire flow of the Orange River is forced through a gap less than 50m wide of hard, well-jointed granitic gneiss, before plunging some 56m into the impressive gorge below.

The Augrabies Falls National Park was proclaimed in 1966, initially to conserve a small area of geological interest around the falls. This is desert wilderness scenery at its best, where the power of running water, over millions of years, has acted on exceptionally hard rock to create one of South Africa's most spectacular geological sites.

Location and getting there

- The falls are situated on the Orange River within the Augrabies Falls National Park, which is about 53,000ha in area and one of the SANParks 'Arid Parks'. You get to this welcome oasis by turning off the N14 at Alheit about 8km west of Kakamas and going another 28km on the tarred R359.

- GPS (Viewsite at falls): 28°35'30"S; 20°20'26"E

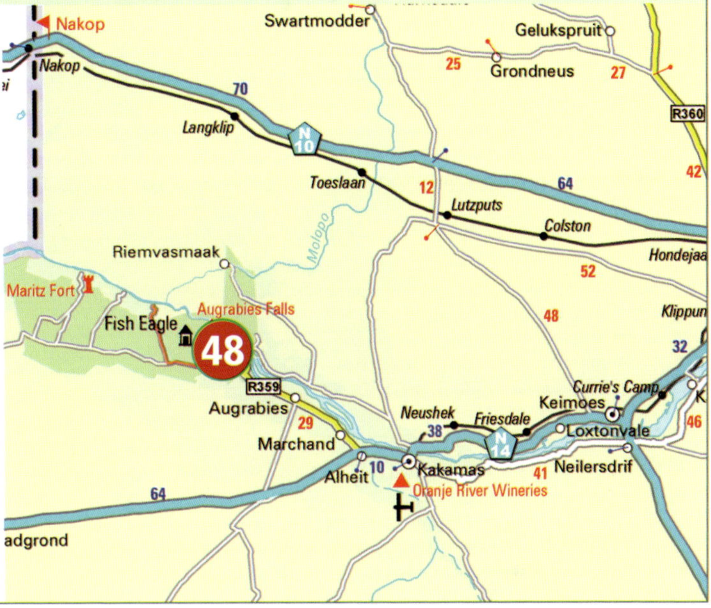

Setting the scene

This is one of South Africa's premier geological attractions, with gneissic crystalline rock exposed almost everywhere and abundant evidence of arid landforming processes, making it a geologist's paradise. At the Main Falls the river narrows dramatically as it drops 56m into an awe-inspiring rocky gorge that continues for another 18km, along the length of which the river falls about 145m. On the northern side of the river, the communal

To the west of the rest camp area, away from the falls and gorge, you find 'Moon Rock', an area of bare, weathered Augrabies granitic gneiss domes showing numerous remnant boulders of eroded gneiss.

South African Tourism

Below the Main Falls, a spectacular view of Augrabies gorge from near Arrow Point, looking downstream towards Oranjekom. The gorge is around 100m deep, up to 500m wide and 18km in length.

Riemvasmaak wilderness desert area is ideal for 4x4 enthusiasts. During the flood season the wide river above the Main Falls spreads out into many side channels to create secondary waterfalls, all plunging into the seething gorge.

Regional geology

The rocks so well exposed here are part of the very extensive **Namaqua Metamorphic Province** that occurs across much of the Northern Cape. This vast and extremely complex geological sequence of metamorphosed, tectonised and intrusive rock formations was formed or reworked during the Namaquan Orogeny around 1,100 million years ago. At Augrabies the underlying geology is part of the **Kakamas Terrain**, possibly much older, and is mostly granitic gneisses of intrusive origin. A series of earthquakes of unknown cause took place here during the period 2010–2012, indicating deep tectonic disturbances.

Landscape and landforms

The Augrabies gorge provides an excellent demonstration of the eroding power of the mighty Orange River, which has, over millions of years, carved out one of the most spectacular gorges in Africa, 18km long, up to 500m wide and about 100m deep. Within the gorge the rocks are largely unweathered, having been recently eroded and smoothed by the river, but further away the landscape is typical of granitic weathering, including exfoliating granite-gneiss domes and residual boulders (see Moon Rock; also Paarl Rock, pp.54–58). The Main Falls is a major knickpoint along the lower part of the Orange River where, under arid conditions, strong headward erosion has cut its way into the planed-down post-African land surface. Above the falls the flat erosion surface that once existed across the entire region is clearly visible, although now considerably dissected. Country-wide uplift less than 5 million years ago lowered

Things to see and do

- **See the Main Falls:** This is the prime attraction, and can be seen from protected viewsites accessed via pathways and boardwalks. The view is particularly spectacular when the river is in flood.
- **Walk to Arrow Point:** This rocky spur, formed by erosion from an earlier river channel, offers superb views both up and down the canyon, and of Twin Falls, and is only a short walk away.
- **Visit Moon Rock:** Large, elongated bare granitic domes composed of Augrabies gneiss and known as 'whalebacks' are located west of the rest camp. Here you can also see enigmatic A-shaped 'pop-up' structures made by exfoliating sheets of granitic gneiss.
- **Drive down-river:** A gravel road along the high ground takes you to Ararat, Oranjekom and eventually to Echo Corner where your call will rebound after a few seconds. The magnificent views from these sites are well worth the drive.
- **Do the hiking trails:** There is a 5km Dassie Trail and the more challenging 3-day Klipspringer Trail that goes into the gorge. On both you are assured of spectacular scenery, desert wildlife and fascinating geology.

the base level of erosion and the early Orange River started cutting back, the gorge lengthening rapidly over time. Gorge formation was assisted by major zones of weakness caused by orthogonal structural jointing; this is revealed by the marked zigzag path that the river has followed. Looking at the bigger picture, the deep incision by the Orange River can be seen as an extraordinary upstream incursion of the western Great Escarpment.

Part of the Oranjekom Complex known as the Swartrante ('black ridges'). This is a small layered mafic intrusion into Riemvasmaak gneiss. The gorge is visible in the background.

Rock types

The underlying geology is fairly simple as almost the entire area is made up mostly of two varieties of granite-gneiss.

- The Augrabies gneiss occurs in the central part of the park and can be clearly seen at the Main Waterfall. It is a medium-grained rock of granitic composition, made of pink-white feldspar, glassy quartz and shiny black mica. It has a mildly gneissic texture and was originally an intrusive granite.
- Surrounding this but found mainly to the northwest is the Riemvasmaak gneiss, which is also considered to be of intrusive origin. This weathers to form characteristic 'pink gneiss', a pinkish-orange rock with a granulitic texture, well exposed in the Oranjekom area.
- Mafic rocks, mainly interlayered gabbro and anorthosite, were intruded later but are little changed, and are known as the Oranjekom Complex. They are seen as blackish ridges in the northwest, known as the Swartrante.

49 Old Koperberg Mine

SOUTH AFRICA'S EARLY MINING HERITAGE

The site of South Africa's first colonial-era copper discovery dating from 1685

Bruce Cairncross; inset: Andrew Hall / CC BY-SA 3.0 Wikimedia Commons

Rugged Koperberg (or Copper Mountain), also known as Van der Stel's copper mine, as seen from the northwestern side, not far off the N14 near Springbok. The circled area showing the fenced-off site is the entrance to Van der Stel's original copper prospecting shaft, '3 fathoms' (about 6 m) in depth. It was declared a National Monument in 1959.

Also called Simon van der Stel's copper mine, this historic but little-visited site was proclaimed a National Monument in 1959. It is an important facet of the proposed Namaqualand Copper Mining Landscape, an extensive region commemorating the origin of South Africa's modern mining industry, based mainly on skills and practice imported from Cornwall.

Location and getting there

- The historic old shaft is located high up on a rocky hill a little more than 1km southeast of the N14, some 6km east of Springbok. The best route is via Carolusberg township and then to follow the signs on the gravel road. On the hillside is a small, fenced-in monument area that limits access to the shaft. On the rockface above, look for the date '1685', engraved some 330 years ago on the green, copper-stained rock face.
- GPS (Monument site): 29°38'58"S; 17°56'14"E

Setting the scene

In 1681 a Namaqua delegation brought specimens of rich copper ore to show to the settlers of the Dutch East India Company at the Cape. This proved to be of considerable interest and several expeditions to find the source area were mounted between 1682 and 1684, but they returned without significant results. In 1685 Simon van der Stel, Governor of the Cape, led a major expedition into Namaqualand to search for copper, leaving Cape Town in August 1685 and reaching the so-called Koperbergen ('Copper Mountains') almost two months later. They carried out widespread prospecting, including digging at least three prospect shafts, and departed in November the same year. Although assay results were encouraging, the remoteness of the area and the difficult terrain, the shortage of water and fuel – and the lack of a suitable harbour – made any thought of mining impossible at that time. Only in 1846 was the first Namaqualand copper-mining enterprise established, leading to a short-lived mining boom (see text box on p.304).

A rather dramatised view of the hilly area around Koperberg, showing copper sites and the camp dating from Simon van der Stel's visit in 1685. The letters 'H.v.R.' apparently mark the Koperberg shaft location.

South African National Archives Repository

Namaqualand's copper riches

In the 1850s copper mining in Namaqualand caused a sensation on the Cape Town Stock Exchange as massive mining speculation took place, and more than 30 small, ultimately unsustainable mining companies were created. In 1852 the first mine to get into production was on the farm Springbokfontein, where the frontier town of Springbok developed. Mines soon opened up at places like Okiep, Nababeep, Spektakel and Concordia, using the skills and practices of Cornish engineers and miners. Ore and later copper metal was transported to the coast, first by ox-wagon down the escarpment to Hondeklip Bay and later by narrow-gauge railway to Port Nolloth. Over the years the mining industry went through several phases of 'boom and bust' in what became known as the Okiep Copper District. In 1937 the Okiep Copper Company was formed and production increased rapidly after the 2nd World War, some 30 mining operations eventually being undertaken in the district. It is ironic and also gratifying that about 1.5km to the east of the original Koperberg discovery, and only in 1971, the enormous Carolusberg Deep copper deposit was discovered by deep drilling, and eventually over 37 million tons of copper ore were produced. Simon van der Stel would have been exceptionally pleased! This was the last of the big mines and by 1998 all the copper mines had been worked out, thus ending a vibrant chapter in South Africa's mining history.

On a hillside directly west of Springbok, the old Blue Mine was the first to produce good copper ore (reportedly 11 tons) in 1852, from what later became known as the rich Okiep Copper District.

Nick Norman

A close-up view of the old Koperberg prospect shaft. Notice the just visible date '1685' (circled) that was engraved 330 years ago into the green, copper-stained face above the shaft entrance.

Originally a granitic rock that intruded around 1,200 million years ago, this coarse-grained rock called 'Nababeep gneiss' is one of the more spectacular rock types found in the Springbok area.

In the Goegap Nature Reserve directly south of Koperberg and east of Springbok, the granitic gneiss terrain comes to life during the springtime wild-flower display.

Regional geology

This region is underlain by the western part of the **Namaqualand Metamorphic Complex**, known as the **Bushmanland Terrain**. It comprises highly metamorphosed granitic gneisses and meta-sedimentary formations, and a range of intrusive granitic and mafic rocks, ranging in age from around 2,000–1,000 million years. The widespread copper mineralisation is associated with the extensive **Koperberg Suite**, which forms countless small intrusive 'basic bodies', often called 'noritoids'. These rocks were originally intruded as dykes into east-northeast-trending structures around 1,060 million years ago, and now occur as discontinuous, pod-like mafic bodies within tight anticlinal structures known as 'steep structures'.

Landscape and landforms

This arid landscape is characterised by numerous rocky koppies that display typical granitic weathering and erosion consisting of rounded boulders, exfoliated domes and intervening sandy plains, as can be seen in the adjoining Goegap Nature Reserve, a wild-flower paradise in springtime.

The hills represent the remains of the elevated Namaqualand Highlands, itself a remnant of an ancient erosion surface that existed above the African land surface.

Rock formations

- The surrounding rock at Koperberg is coarse-grained Nababeep gneiss and the copper-bearing rock is intrusive noritoid.
- During the 1685 expedition the prospect shaft was sunk to a depth of '3 fathoms' (about 6m) into what we now name a copper-bearing 'noritoid' within a 'steep structure'.
- The mining engineer who accompanied the 1685 expedition wrote this: 'I am entirely convinced of the favourable character of these mines; in the event of mining operations being continued, richer and richer, and better minerals will be found, for the hills concerned extend several miles in length and breadth, and hold minerals almost everywhere, and in some places water for man and beast is plentiful, as the Commander found on personal inspection'.
- These are prophetic words but it took over 250 years for the full potential to be realised.

Things to see and do (most of these are Provincial Heritage Sites)

- **Visit the old Koperberg prospect shaft:** Try to picture the excitement of the Dutch prospectors digging here in 1685 in what was to them a very foreign land.
- **See the old Blue Mine in Springbok:** This was the first producing mine in the district, dating from 1852, when 11 tons of copper ore were produced. The open mine overlooks the western side of the town.
- **Visit the mining museum at Nababeep:** This small museum illustrates the local copper industry and has many mining artefacts, including the narrow-gauge steam locomotive called Clara.
- **See the old mine buildings at Okiep:** In the 1870s this was ranked as the world's richest copper mine. The Cornish miners left a legacy in the form of the well-preserved beam pump house and the chimney (smoke stack) built in 1880, both typical of those dotted about the Cornish landscape.
- **See Orbicule Hill at Concordia:** A small hill to the west of the village has a magnificent display of rare orbicular diorite formed by the segregation of two granitic magmas, one forming the concentric orbicules, and the other the matrix.

Historic National Monuments at the village of Okiep that hark back to the days of the pioneer Cornish miners and engineers: (left) a well-preserved typical beam pump and its stone-built housing; (below) a smelter smoke stack that was once part of the local copper-mining industry.

'Clara' (now at the Nababeep Mining Museum) is a small, narrow-gauge steam engine that once transported rich copper exports by railway line from Springbok to the harbour at Port Nolloth.

50 Fraserburg Fossil Surface

A LANDSCAPE PRESERVED IN TIME

An exceptionally well-preserved Karoo palaeo-surface, famous for its reptilian tracks

The fenced-off fossil surface is found on the farm Gansfontein, 4km northwest of Fraserburg. One's first view – of typical semi-desert farmland of the Upper Karoo – does not impress. But a guided tour and detailed inspection will reveal fascinating evidence of life that lived around here 260–255 million years ago, like the inset fossil trackway (wetted for effect).

Fossilised mammal-like reptilian footprints and trace fossils preserved on an ancient land surface near Fraserburg reveal the life that existed here during the Permian, together with a range of primary sedimentary structures shown in fine detail. This palaeo-surface is a 'snapshot in time' that provides a rare glimpse of what took place during deposition of the Karoo's Beaufort sediments around 260–255 million years ago, when four-legged, five-toed dinocephalians squelched their way through muddy pools and shores, leaving lines of tracks.

■ Location and getting there

- Fraserburg is a quiet farming town well off the beaten track in the southwestern Upper Karoo. Several gravel roads lead there but the best route is the tarred R353 from the N1, turning off at Leeu-Gamka and passing over the scenic Nuweveld Escarpment via the Teekloof Pass. The fossil surface is located on the farm Gansfontein about 4km northwest of the town and you can only visit the site in the company of an accredited local guide.

- GPS (fossil surface): 31°54′08″S; 21°28′54″E

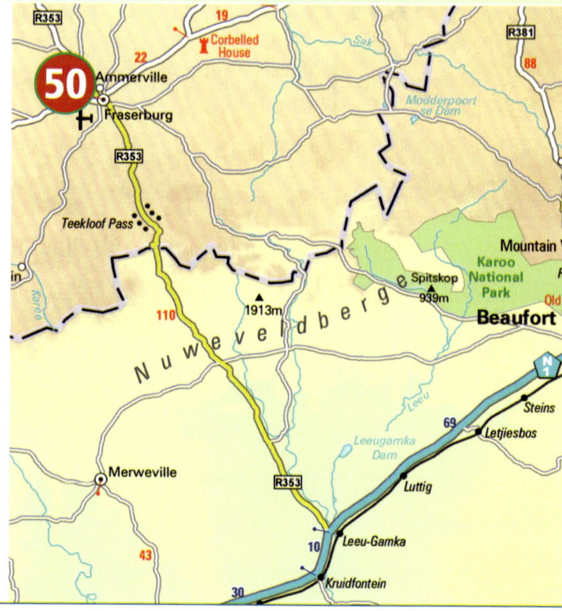

The 'blue plaque', erected by the Geological Society of South Africa in 1989 at this remote and little-visited geological site, provides a summary of the local palaeontological spectacle.

South of Fraserburg the thickness and make-up of the sedimentary Beaufort Group (mostly shale) can best be appreciated at the Teekloof Pass that makes its way down the Nuweveld Escarpment from the Upper Karoo plateau to the plains of the Great Karoo.

Setting the scene

The fossil surface, about 60m long and 25m wide, was discovered by a local farmer after flood waters washed away overburden and exposed the remarkable underlying rock floor. In 1986 Coenie de Beer, a Geological Survey geologist mapping the area, described the site and took up the cause for its conservation. An information pamphlet was compiled and the Geological Society of South Africa erected a 'blue plaque'. But despite its importance and outstanding geoheritage value, the site still has no formal protection status.

Regional geology

- The area is underlain by sedimentary rocks – mainly mudstones and lesser sandy siltstones – of the Permian **Beaufort Group.**
- This is an enormously thick package of continental flood plain sediments of the **Karoo Supergroup** that are here undeformed and virtually flat-lying.
- In the modern biostratigraphic classification of the Karoo sequence, the area falls within the **Tapinocephalus Assemblage Zone**, characterised by a variety of dinocephalian (mammal-like reptilian) fossils.

The amazing story of the Permian fossil surface

- There are no fossil bones preserved in the mud, indicating that no animals actually died here – what you see are trace fossils, the signs left by creatures moving across the muddy site.
- Periodic floods deposited layers of silt, and resulted in short-lived muddy pools that provided the conditions for the capture of the animal tracks and sedimentary structures.
- Large four-legged, five-toed dinocephalians, probably plant-eaters, squelched their way along the muddy shores and their lumbering reptilian gait left characteristic lines of tracks.
- A single set of larger stubby-toed footprints was probably made by a solitary *Bradysaurus*; there are also marks of dragging tails and there are other much smaller footprints, delicate worm tracks, and arthropod and fish-fin trails.
- There's evidence of mud splatters, and ripples that formed as a result of wind.
- These trace fossils give us a glimpse into what took place on the prehistoric Karoo flood plain over a short period of time. The flood plain would have dried rapidly in the prevailing arid conditions, the mud hardened and polygonal mud-cracks developed.
- The next flood across the flood plain would bring in another layer of silt that was deposited over the earlier, semi-hardened surface, preserving it forever.
- This process of alluvial deposition continued for millions of years, building up layer upon layer, until eventually the sediment was deeply buried and became hardened to beds of sandstone, siltstone and mudstone of the Beaufort Group.
- A recent Karoo flood, albeit on a much smaller scale, serendipitously re-exposed the fossil surface. Unfortunately, the exposed fossil surface is now deteriorating as a result of natural weathering.

Run-off surface drainage channel with lingoid ripples and fine mud cracks.

Mud splatters (drip marks) probably from mud falling off a large muddy reptile, much as it does from elephants in mud pools today.

Small ripple marks, probably wind generated; several different sets are evident in different directions, formed at different times.

A large (about 30cm across) unidentified reptilian footprint on an expanse of mud-cracked mudstone – and an indication of the semi-arid climate.

Landscape and landforms

Today this area is a typical undulating plain of the Upper Karoo, with higher ground marking the presence of Karoo sills or dykes. Drainage lines are branched, or dendritic, and generally flow northwards. To the south of Fraserburg lies the rugged Nuweveld Escarpment, part of South Africa's country-wide Great Escarpment, and here dominated by the presence of thick Karoo dolerite sills.

Rock formations and trace fossils

- The exposed fossil surface shows horizontal, interlayered beds of sandy siltstone and mudstone that are part of the mudstone-dominated Abrahamskraal Formation.
- At the time of deposition the Karoo Basin was a vast alluvial flood plain crossed by meandering ephemeral rivers and was prone to periodic sediment-laden flooding.
- The climate was semi-arid but there was probably considerable riverine vegetation, and a wide diversity of reptilian species must have thrived here.
- Now preserved within a layer of thinly bedded greenish siltstone is an original Karoo land surface where early vertebrate life walked 260–255 million years ago, leaving their impressive footprints. There are also delicate fish-fin trails, and arthropod and worm trails.
- Many fascinating primary sedimentary structures reveal what took place during flooding and drying-out of the surface. Structures preserved in the mud include mud pellets, run-off channels, falling water marks, ripple marks, rill marks and mud cracks (see images opposite).

Things to see and do

- **Visit Gansfontein:** It is on private land and can only be visited with an accredited guide by arrangement with the Fraserburg Museum. Exercise care when walking across fragile places and look for both animal tracks and sedimentary structures, then let your mind wander back to a time when the Karoo was a very different place.
- **Visit the museum:** This is housed in the historic Old Parsonage and has a small collection of Karoo fossils. Another good Karoo fossil display can be found at the Karoo National Park near Beaufort West.
- **Iziko South African Museum:** If you can't get to Fraserburg there is an excellent exhibit of Karoo fossils at the museum in Cape Town. A well-illustrated exhibit on the Fraserburg fossil surface forms part of the Stone Bones display.

The Old Parsonage building, now the site of Fraserburg's Museum, which has a small Karoo fossil section.

Glossary

Aeolian – relating to or arising from the action of the wind; fine-grained sandstone thus formed is called 'aeolianite' or 'dune rock'.

African erosion surface – a long-lived, probably multicyclic land surface of deep weathering, erosion and planation that developed on the relatively stable southern African subcontinent after the break-up of Gondwana.

Alluvial – comprised of loose, sedimentary material of variable origin that is moved and deposited by running water.

Alluvial fan – a roughly triangular deposit of sedimentary debris accumulated at the base of a steep mountain slope, assisted by gravity.

Amygdales – small rounded or pipe-like, mineral-filled cavities found in lava, and which were once steam- or gas-filled bubbles. A lava containing these is called 'amygdaloidal'.

Anticlinorium – a large-scale, broad fold structure having an overall arch-like shape, made of a number of smaller superimposed synclines and anticlines.

Bankenveld – an asymmetrical ridge and step-like topography resulting from preferential erosion of gently dipping softer and harder sedimentary strata. Also known as 'cuesta' landscape.

Base level – the lowest level to which a river can erode under stable Earth conditions; and which will change over geological time.

Basement – a general term for the oldest 'floor rocks' in southern Africa, of granitic, gneissic and greenstone types, and upon which younger rock formations were deposited.

Basin – a depressed region of the crust which if at low level or below sea level, could become a depositary for new sediment, later to become sedimentary rocks.

Batholith – a large, cross-cutting intrusive of plutonic rock, now exposed by erosion; typical of granitic intrusions and generally more than 100km^2 in area (see also 'pluton').

Bedding – the very varied primary layering of sedimentary rocks that form defined beds or strata, which provide evidence of a wide variety of depositional conditions.

Braided – describes the flow of low-gradient river or stream channels that split and re-join repeatedly across a flood plain, and are later reflected by the geology.

Breccia – rock made up of angular fragments, usually of differing size, within a finer-grained matrix. Different breccia types exist, e.g. cave breccia, fault breccia, volcanic breccia or scree breccia.

Butte – a small, isolated hill with steep sides, and probably a remnant of a mesa, e.g. Spandau Kop at Graaff-Reinet.

Cave breccia – secondary deposits that have accumulated in some caves from heterogeneous debris that was washed in and cemented by limestone into a concrete-like mass. It is a rich source of many types of fossil.

Cave formations – finely banded secondary calcite (called 'travertine') that develops within limestone or dolomite caves by dripping and evaporation of carbonate-rich ground water, and includes unique flowstones, stalactites and stalagmites.

Clast – a rock fragment, pebble or mineral particle of any size or shape that makes up a sedimentary rock. A rock made of clasts is known as a 'clastic rock'.

Craton – a very large mass of early Earth's crust, created by the intrusion of granitic rocks into more primitive greenstone rocks, and which has remained relatively stable ever since the Archaean.

Cross-bedding – a primary sedimentary structure in which the initial deposition by water or wind was rapid, and on an inclined surface.

Crust – the thin outermost skin of the Earth above the upper mantle, varying markedly in thickness from more than 35km beneath continental areas to around 7km beneath oceanic areas.

Deformation – the varied processes whereby rock formations are disrupted by tectonic forces of compression or extension within the crust, resulting in faulting, rifting, folding, thrusting and subduction.

Detritus – material of any size and type, derived by weathering and erosion of rocks.

Diagenesis – part of the process of lithification (rock formation) whereby deposited sediment is changed into sedimentary rock at relatively low pressures and temperatures.

Dip – the maximum angle of inclination of a sedimentary bed from the horizontal.

Dip-slope – the angle of slope of the ground surface which closely corresponds to the dip of the underlying strata. The opposing, steeper scarp-like rock face is known as the 'scarp-slope'.

Dome – where all sedimentary beds dip outwards to form a roughly circular or oval pattern due to central uplift. Erosion of a domal feature may expose the inner rock formations, e.g. the Johannesburg Dome in Gauteng.

Dune rock – see 'aeolian'.

Dyke – igneous rock that forms a near-vertical elongated intrusion. They vary widely in width, may continue for kilometres, and are connected to a deeper magma source.

Erosion – the natural landscaping process that results in the removal and transportation of weathered rock material and soil, by the action of gravity, water, waves and wind.

Erosion surface – an older land surface that formed by long-lived erosion and planation during a period of Earth stability.

Escarpment – an elongated cliff-like face or very steep slope which is being actively eroded.

Exfoliation – concentric rock weathering, on any scale, in which curved surfaces peel off from the core of the rock; also called 'onion-skin weathering'.

Extrusion – an igneous rock that has erupted and solidified on the Earth's surface, usually as a lava or similar volcanic rock.

Fabric – the spatial arrangement or alignment of minerals that make up the texture of an igneous or metamorphic rock.

Felsic – first of the four-fold classification of igneous rocks, based on high silica content and usually indicated by colour; the name is derived from its main constituents, feldspar and silica (see Table 2, p.16).

Fluvial – rocks formed by very shallow-water deposition, by rivers or streams on a flood plain; typically sandstones, grits and conglomerates.

Foliation – characteristic texture of many metamorphic rocks showing a preferred alignment of certain minerals, such as mica in schists.

Formation – a formal division of stratigraphy used to classify defined geological units that formed under similar depositional conditions within a basin. A number of related Formations make up a formal Group.

Geological Timescale – the internationally accepted and complex division of rock formations worldwide, and their geological ages (see Table 6, p.24).

Gondwana – an ancient landmass or supercontinent that once existed, and which broke up by large-scale crustal fracturing and the movement of crustal plates over an extended period of geological time, starting more than 140 million years ago, to form today's southern continents, Madagascar and India.

Greenstone – strongly folded, metamorphosed and often schistose rocks that make up Earth's oldest geological formations; often characterised by containing greenish minerals such as serpentine, amphiboles and chlorite.

Group – an important division of stratigraphy that is used to unite different geological Formations occurring within a common depositional basin or metamorphic terrain; may be part of an even larger grouping called a 'Supergroup'.

Hydrothermal – relating to or produced by very hot, natural water systems within the Earth's crust and the resulting changes to rocks; sulphide minerals and gold are often deposited by this process.

Hypabyssal – one of three types of igneous rock, this type having crystallised at relatively shallow depth below surface so that it's coarser-grained than volcanic rock but finer-grained than plutonic rock (see Table 2, p.16).

Igneous – having solidified directly from the crystallisation of molten magma, of varying composition, and in a wide variety of forms. They include all volcanic, hypabyssal and plutonic rock types.

Inlier – a window-like exposure of older rocks that are completely surrounded by younger formations, formed either by a domal structure or by erosion of the younger rocks.

Intermediate – second of the four-part classification of igneous rocks, based on having a lower silica content and often indicated by a darker colour (see Table 2, p.16).

Intrusion – a variety of igneous processes whereby magma is forcibly emplaced upwards into deep fractures and overlying strata within the Earth's crust. Such rocks, which include dykes and sills, are collectively known as 'intrusives'.

Island arc – a usually curved belt of volcanic islands that has developed above a zone of crustal subduction, and is related to the movement of oceanic crustal plates, e.g. the islands of Japan.

Joints – natural cracks or fracture planes within a rock mass on which no movement can be seen. They are fundamental factors in the weathering and erosion of rock.

Juvenile – the early part of the cycle of erosion, describing a youthful river system that is actively eroding, and is characterised by steep valleys and high stream gradients.

Kaapvaal Craton – the ancient and relatively stable core to the southern African continent that formed before 3,000 million years ago, from the consolidation of greenstone belts and large intrusive granitic batholiths.

Karst – development within existing limestone and dolomite rocks that often form rough topography and extensive cave systems, following large-scale dissolution of carbonate rock by ground water, along joints, fractures and faults.

Land surface – at least four cycles of land-surface erosion have been identified in South Africa, of which the African surface following the break-up of Gondwana was the longest and most extensive (see Table 7, p.37).

Lapilli – small molten lava fragments blasted out of a volcano and solidified in the air to form rounded aggregates that fall to Earth; usually of fine- to medium-grain size.

Layering – pattern of deposition of extruded lava; or beds in a sedimentary sequence; or igneous rocks that crystallise very slowly as distinct igneous layers.

Lithification – the complex chemical and physical processes that take place during deep burial of sediment, and the conversion of unconsolidated sediment into sedimentary rock.

Lopolith – a large, basin-shaped igneous intrusion that forms within the enclosing rock strata, and may show igneous layering.

Mafic – third of the four-part classification of igneous rocks, based on having a relatively low silica content and invariably indicated by a dark colour; the name is derived from its higher content of magnesium and ferric iron (see Table 2, p.16).

Mantle – the very wide zone between the Earth's crust and the solid inner core; it is in the Earth's upper mantle below the crust that mobile, upward-moving magmas are first generated.

Massive – having a homogeneous structure without obvious sedimentary bedding, igneous layering or rock jointing.

Matrix – the fine-grained groundmass to a porphyritic igneous rock; or the finer-grained material between larger fragments in sedimentary rocks such as conglomerate.

Mature – the middle part of the cycle of erosion, describing a river system in a moderately to slowly eroding landscape, where river gradients are relatively low.

Mesa – a large, flat-topped 'table mountain', typically steep-sided and usually isolated, e.g. Platberg near Harrismith.

Metamorphic – descriptive of a type of rock that has been subjected to considerable heat and/or pressure so that the original minerals may have been recrystallised into new minerals and textures (see Table 5, p.20).

Minerals – the common building blocks of all rocks (see Table 1, pp.14–15).

Nepheline – a rare, sodium-rich silicate mineral associated only with syenites and foyaites.

Orogeny – a large-scale mountain-building event that originates from a subducting crustal plate, and results in compressive tectonic activity within the Earth's crust, e.g. the Cape Fold Belt.

Orthogneiss – coarse-grained rock that is formed from original igneous rocks, usually of plutonic origin, during orogenesis and high-grade metamorphism.

Outlier – isolated occurrence of younger rocks that is completely surrounded by older formations, due to erosion.

Paragneiss – coarse-grained rock that is formed from original sedimentary rocks during orogenesis and high-grade metamorphism.

Peneplain – extensive planed-down land surface, generally flat to very gently sloping, a result of prolonged crustal stability and sustained weathering and erosion.

Phenocrysts – larger crystals of earlier-formed minerals, such as quartz or feldspar, within a much finer-grained groundmass; a rock containing phenocrysts is called 'porphyritic'.

Pillow lava – usually mafic lava that is erupted under water, cools very rapidly to form a glassy outer surface, and accumulates as piled-up, pillow-like masses below sea level.

Pipe – a carrot-shaped volcanic intrusion (or diatreme) that is formed by the rapid and violent eruption of gas-rich magma through surrounding wall rocks.

Pluton – a relatively smaller plutonic mass, generally rounded or oval in outline, of coarse-grained igneous rock, now exposed by erosion, less than 100km^2 in area, and typical of smaller granitic intrusions (see also 'batholith').

Plutonic – one of three types of igneous rock, this type having been intruded and then cooled at considerable depth, where it crystallised slowly, forming coarse-grained rocks (see Table 2, p.16).

Porphyry – hypabyssal or volcanic igneous rock that contains large, earlier-formed crystals called 'phenocrysts' within a finer-grained matrix.

Pseudotachylite breccia – a rare, shock-generated brecciated rock type typically associated with meteorite impact structures, such as at the Vredefort Dome.

Red bed – usually a sandstone with a reddish coloration, caused by the pervasive presence of early-formed iron oxides (see text box, p.243).

Regression – a period when sea level was receding, young deposits were exposed to the atmosphere and to erosion, and the coastline became more extensive.

Ring dyke – strongly curved or semi-circular hypabyssal intrusion filling circular faults around a sunken area such as a volcanic caldera.

Rock Cycle – see Figure 3, p.25.

Scarp-slope – see 'dip-slope'.

Scree – weathered and eroded rock debris, which usually collects at the base of steep slopes or scarps; also called 'talus'.

Sediment – unconsolidated rock and mineral material derived from erosion, and ranging in size from boulder and pebbles to very fine-grained suspended mud and clay.

Sedimentary – type of rocks of clastic, chemical and organic origin, mostly deposited as sediment by water, but also formed by the action of wind and ice (see Table 3, p.18 and Table 4, p.19).

Senile – descriptive of a river system that is old, the final stage in the cycle of erosion, where the landscape has been completely worn down by erosion to produce a flood plain and meandering rivers.

Shear zone – narrow zone of movement that has experienced considerable stress and deformation at different levels within the crust.

Sill – a sheet-like or tabular hypabyssal igneous rock that intrudes a sequence of older rocks, either parallel to the bedding (concordant) or at an inclination to it (discordant).

Stalactite – a cave formation of travertine that has grown downwards from a cave roof by dripping and evaporation of calcium-rich water.

Stalagmite – a pillar-like formation of travertine that has grown upwards from a cave floor by dripping and evaporation of calcium-rich water.

Strata – non-specific name for layers of sedimentary rock, which may show individual characteristics.

Stratigraphy – the detailed description, nomenclature and correlation of all rock strata, including igneous and metamorphic rocks, and their sequence in time.

Strike – used in the mapping of sedimentary rocks, measured horizontally at 90° to the direction of maximum dip.

Stromatolite – a mound-like, finely laminated algal structure of very variable size that reflects the fossil evidence of primitive sediment-collecting micro-organisms called 'cyanobacteria'.

Subduction (zone) – a large-scale process whereby one crustal plate is slowly but irresistibly forced under a neighbouring crustal plate by the force of plate tectonics, and which causes orogenic deformation and generation of new magmas.

Supergroup – the uppermost hierarchical level in stratigraphic classification, representing the largest assemblage of mutually similar geological Groups.

Synclinorium – a large-scale, broad, composite fold structure with an overall basin-like shape and made of a series of smaller superimposed synclines and anticlines.

Tectonic – pertaining to the deformational forces and outcomes of the broad-scale structure and movement of the Earth's crust.

Terrain – an area having common physical features, or referring to a broad regional geological setting.

Thrusting – within a fold belt, the process whereby a discrete zone of very low-angle faulting is developed, caused by considerable compression within a highly folded sequence.

Transgression – a period when sea level was rising, causing drowning of the coastline and sedimentary deposition moved further inland.

Travertine – a secondary form of finely banded calcite that makes cave formations.

Ultramafic rocks – fourth of the four-part classification of igneous rocks, based on very low silica content, and invariably dark in colour (see Table 2, p.16).

Unconformity – a change in the structural attitude between two different geological formations that are in contact. It indicates a break in sedimentary deposition, perhaps of considerable time and in different conditions.

Uplift – the slow elevation by internal tectonic forces of part of the Earth's crust relative to sea level.

Volcanic – descriptive of one of three types of igneous rock, this being fine-grained to glassy rock that erupted on the Earth's surface, or even under water (see Table 2, p.16).

Weathering – the chemical, biological and mechanical changes that break down all surface rocks, dependent on climate and assisted by the degree of rock jointing.

Further reading

There is a wide range of well-known and generally accessible books covering a variety of geological and geomorphological topics in South Africa. A good selection of these is listed below, alphabetically by first author or by title. They vary in scientific content from fairly academic to very easy-to-read, and include useful South African reference books.

Bulpin, TV. 1985. *Scenic Wonders of Southern Africa*. Books of Africa, Muizenberg.

Cairncross, B. 2004. *Field Guide to Rocks and Minerals of Southern Africa*. Struik Publishers, Cape Town.

Cairncross, B & Dixon, R. 1995. *Minerals of South Africa*. The Geological Society of South Africa, Johannesburg.

Compton, JS. 2004. *The Rocks & Mountains of Cape Town*. Double Storey Books, Cape Town.

Coyne, P. 2010. *A Guide to South Africa's Mountain Passes and Poorts*. Osborne Porter Literary Services, Westville.

Curror, WD. *Golden Memories of Barberton*. (Revised and enlarged by Hans Bornman.) African Pioneer Mining, Barberton.

Eales, HV. 2001. *A First Introduction to the Geology of the Bushveld Complex*. Council for Geoscience, Popular Geoscience Series #2.

Eales, H. 2007. *Riddles in Stone: Controversies, Theories and Myths about Southern Africa's Geological Past*. Wits University Press, Johannesburg.

Grab, S & Knight, J (editors). 2014. *Landscapes and Landforms of South Africa*. Springer-Verlag, Berlin-Heidelberg (in press).

Hilton-Barber, B & Berger, LR. 2002. *Field Guide to the Cradle of Humankind: Sterkfontein, Swartkrans, Kromdraai & Environs World Heritage Site*. Struik Publishers, Cape Town.

Holmes, P & Meadows, M (editors). 2012. *Southern African Geomorphology. Recent Trends and New Directions*. Sun Media, Bloemfontein.

Illustrated Guide to South Africa. (Original text by TV Bulpin.) 1978. Reader's Digest Association of South Africa, Cape Town.

Johnson, MR, Anhaeusser, CR & Thomas, RJ (editors). 2006. *The Geology of South Africa*. The Geological Society of South Africa, Johannesburg & the Council for Geoscience, Pretoria.

King, LE. 1951. *South African Scenery: a Textbook of Geomorphology*, 2nd edition. Oliver & Boyd, London.

MacRae, C. 1999. *Life Etched in Stone: Fossils of South Africa*. The Geological Society of South Africa, Johannesburg.

McCarthy, T. 2009. *How on Earth? Answers to the Puzzles of our Planet*. Struik Nature, Cape Town.

McCarthy, T & Rubidge, B. 2005. *The Story of Earth & Life: A Southern African Perspective on a 4.6-billion-year Journey*. Struik Publishers, Cape Town.

Norman, N. 2013. *Geology off the Beaten Track: Exploring South Africa's Hidden Treasures*. Struik Nature, Cape Town.

Norman, N & Whitfield, G. 2006. *Geological Journeys: a Traveller's Guide to South Africa's Rocks and Landforms*. Struik Nature, Cape Town.

Reimold, WU, Brandt, D, De Jong, R & Hancox, J. 1999. *Tswaing Meteorite Crater: an Introduction to the Natural and Cultural History of the Tswaing Region including a Description of the Hiking Trail*. Council for Geoscience, Popular Geoscience Series #1.

Reimold, WU & Gibson, RL. 2005. *Meteorite Impact! The Danger from Space*. Chris van Rensburg Publications, Johannesburg.

Shoreline – Discovering South Africa's Coast. 2010. (Based on SABC2 documentary TV series.) Struik Travel & Heritage, Cape Town.

Viljoen, MJ & Reimold, WU. 1999. *An Introduction to South Africa's Geological and Mining Heritage*. Mintek in association with the Geological Society of South Africa, Johannesburg.

Wannenburgh, A & Dickson, JR. 1987. *The Natural Wonder of Southern Africa*, 2nd edition. C. Struik, Cape Town.

From the Sani Pass Hotel, a view of the southern Drakensberg, looking up the Mkhomazana River valley, towards the Sani Pass

Index

D

E

F

G

H

T

U

V

W

Z